T0215859

Frontiers in Mathematics

Marko Lindner

Infinite Matrices
and their
Finite Sections

An Introduction to the
Limit Operator Method

Birkhäuser Verlag
Basel · Boston · Berlin

Author's address:

Marko Lindner
Department of Mathematics   and   Faculty of Mathematics
University of Reading   Chemnitz University of Technology
Whiteknights, PO Box 220   09107 Chemnitz
Reading, RG6 6AX   Germany
UK
e-mail: m.lindner@reading.ac.uk

2000 Mathematical Subject Classification, Primary: 47B38, 47L10, 65J10, 65R20;
Secondary: 46E40, 47B10, 47B35, 47C05, 47L15, 47L20, 65F15, 74J20

A CIP catalogue record for this book is available from the
Library of Congress, Washington D.C., USA

Bibliographic information published by Die Deutsche Bibliothek
Die Deutsche Bibliothek lists this publication in the Deutsche Nationalbibliografie;
detailed bibliographic data is available in the Internet at <http://dnb.ddb.de>.

ISBN 3-7643-7766-6 Birkhäuser Verlag, Basel – Boston – Berlin

© 2006 Birkhäuser Verlag, P.O. Box 133, CH-4010 Basel, Switzerland
Part of Springer Science+Business Media
Cover design: Birgit Blohmann, Zürich, Switzerland
Printed on acid-free paper produced from chlorine-free pulp. TCF ∞

ISBN 3-7643-7766-6   ISBN 3-7643-7767-4 (eBook)
ISBN 978-3-7643-7766-3

9 8 7 6 5 4 3 2 1   www.birkhauser.ch

To Diana

Introduction

In this book we are concerned with the study of a certain class of infinite matrices and two important properties of them: their *Fredholmness* and the *stability* of the approximation by their finite truncations. Let us take these two properties as a starting point for the big picture that shall be presented in what follows.

Fredholmness   Stability

We think of our infinite matrices as bounded linear operators on a Banach space E of two-sided infinite sequences. Probably the simplest case to start with is the space $E = \ell^2$ of all complex-valued sequences $u = (u_m)_{m=-\infty}^{+\infty}$ for which $|u_m|^2$ is summable over $m \in \mathbb{Z}$.

The class of operators we are interested in consists of those bounded and linear operators on E which can be approximated in the operator norm by band matrices. We refer to them as *band-dominated operators*. Of course, these considerations are not limited to the space $E = \ell^2$. We will widen the selection of the underlying space E in three directions:

- We pass to the classical sequence spaces ℓ^p with $1 \leq p \leq \infty$.
- Our elements $u = (u_m) \in E$ have indices $m \in \mathbb{Z}^n$ rather than just $m \in \mathbb{Z}$.
- We allow values u_m in an arbitrary fixed Banach space $\mathbf{X}$ rather than $\mathbb{C}$.

So the space E we are dealing with is characterized by the parameters $p \in [1, \infty]$, $n \in \mathbb{N}$ and the Banach space $\mathbf{X}$; it will be denoted by $\ell^p(\mathbb{Z}^n, \mathbf{X})$. Note that this variety of spaces E includes all classical Lebesgue spaces $L^p(\mathbb{R}^n)$ with $1 \leq p \leq \infty$ if we identify a function $f \in L^p(\mathbb{R}^n)$ with the sequence of its restrictions to the cubes $m + [0, 1)^n$ for $m \in \mathbb{Z}^n$, understood as elements of $\mathbf{X} = L^p([0, 1)^n)$.

For our infinite matrices $[a_{ij}]$ acting on $E = \ell^p(\mathbb{Z}^n, \mathbf{X})$, the indices i, j are now in $\mathbb{Z}^n$, and the entries a_{ij} are linear operators on $\mathbf{X}$. Clearly, such band-dominated operators can be found in countless fields of mathematics and physics. Just to mention a few examples, we find them in wave scattering and propagation problems [21], quantum mechanics [38], signal processing [65], small-world networks [52], and biophysical neural networks [11]. Prominent examples are convolution-type operators, Schrödinger (for example, Almost Mathieu) operators, Jacobi operators and other discretizations of partial differential and pseudo-differential equations.

Stability of the Approximation by Finite Truncations

If a bounded and linear operator on E, generated by an infinite matrix A, is invertible, then, for every right-hand side $b \in E$, the equation

$$Au = b \qquad (1)$$

has a unique solution $u \in E$. To find this solution, one often replaces equation (1) by the sequence of finite matrix-vector equations

$$A_m u_m = b_m, \qquad m = 1, 2, \ldots \qquad (2)$$

where $A_m = [a_{ij}]_{|i|,|j| \leq m}$ is the so-called mth finite section of the infinite matrix A and b_m is the respective finite subvector of the right-hand side b.

The naive but often successful idea behind this procedure is to "keep fingers crossed" that (2) is uniquely solvable – at least for all sufficiently large m – and that the solutions u_m of (2) componentwise tend to the solution u of (1) as m goes to infinity. One can show that this is the case for every right-hand side $b \in E$ if and only if A is invertible and (A_m) is *stable*, the latter meaning that all matrices A_m with a sufficiently large index m are invertible and their inverses are uniformly bounded.

Fredholm Operators

If a bounded and linear operator A on E is not invertible, then it is not injective or not surjective; that is, either

$$\ker A := \{u \in E : Au = 0\} \neq \{0\} \qquad \text{or} \qquad \operatorname{im} A := \{Au : u \in E\} \neq E,$$

or both. As an indication of how badly injectivity and surjectivity are violated, one looks at the dimension of $\ker A$ and the co-dimension of $\operatorname{im} A$ by defining the integers $\alpha := \dim \ker A$ and $\beta := \dim(E/\operatorname{im} A)$, provided $\operatorname{im} A$ is closed. The operator A is called a *Fredholm operator* if its image $\operatorname{im} A$ is closed and both α and β are finite. In this case, the integer $\alpha - \beta$ is called the *Fredholm index* of A.

So, if A is a Fredholm operator, then A, while not necessarily being invertible, is still reasonably well-behaved in the sense that the equation (1) is solvable for all right-hand sides b in a closed subspace of finite co-dimension, and the solution u is unique up to perturbations in a finite-dimensional space, namely $\ker A$.

As stated earlier, we are going to study these two properties, the stability of the finite section approximation and the Fredholm property, for our infinite matrices alias band-dominated operators. To do this we shall introduce a third property, called *invertibility at infinity*, that is closely related to both Fredholmness and stability, and we present a tool for its study: the method of *limit operators*.

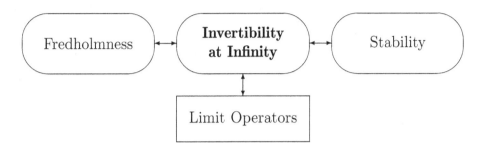

Invertibility at Infinity

A band-dominated operator A is said to be *invertible at infinity* if there are two band-dominated operators B and C and an integer m such that

$$Q_m AB = Q_m = CAQ_m \qquad (3)$$

holds, where Q_m is the operator of multiplication by the function that is 0 in and 1 outside the discrete cube $\{-m, m\}^n$.

This property is intimately related with Fredholmness on $E = \ell^p(\mathbb{Z}^n, \mathbf{X})$. Indeed, if $1 < p < \infty$, then Fredholmness implies invertibility at infinity whereas the implication holds the other way round if $\mathbf{X}$ is a finite-dimensional space. Thus both properties coincide if $E = \ell^2$, for example.

Our other main issue is that concerning the stability of the sequence (A_m) in (2). One easily reduces this problem to an associated invertibility at infinity problem. Instead of the sequence of finite matrices $A_1, A_2, \ldots$ we look at the infinite block diagonal matrix

$$A' := \operatorname{diag}(A_1, A_2, \ldots).$$

The sequence (A_m) turns out to be stable if and only if A' is invertible at infinity.

We will also present a slightly more involved method of assembling a sequence (A_m) of operators to one operator A' by increasing the dimension of the problem from n to $n + 1$. Roughly speaking, we stack infinitely many copies of the space E, together with the operators A_m acting on them, into the $(n + 1)$th dimension. With this stacking idea we can also study the stability of approximations by infinite matrices A_m.

Limit Operators

To get an idea of how to study invertibility at infinity, we think of a band-dominated operator A on E as an infinite matrix $[a_{ij}]$ again. For every $m \in \mathbb{N}$, the operators $Q_m A$ and $A Q_m$ in (3), and hence the invertibility at infinity of A, are independent of the matrix entries a_{ij} with $i, j \in \{-m, m\}^n$. Consequently, all information about invertibility at infinity is hidden in the asymptotics of the entries a_{ij} towards infinity.

For band-dominated operators, the only interesting directions for the study of these asymptotics are the parallels to the main diagonal since the matrix entries decay to zero in all other directions.

For our journey along the diagonal, we choose a sequence $h = (h_m) \subset \mathbb{Z}^n$ tending to infinity and observe the sequence of matrices $[a_{i+h_m, j+h_m}]$ as $m \to \infty$. If this sequence of matrices, alias operators on E, converges in a certain sense, then we denote its limit by A_h and call it the *limit operator* of A with respect to the sequence h.

We call A a *rich* operator if it has sufficiently many limit operators in the sense that every sequence h tending to infinity has a subsequence g such that A_g exists. In this case, all behaviour of A at infinity is accurately stored in the collection of all limit operators of A. We denote this set by $\sigma^{\mathrm{op}}(A)$ and refer to it as the *operator spectrum* of A. For operators $A \in \mathrm{BDO}^p_{\mathcal{S},\$}$; that is the set of all rich band-dominated operators with the additional technical requirement that A is the adjoint of another operator if $p = \infty$, we prove the following nice theorem.

Theorem 1. *An operator* $A \in \mathrm{BDO}^p_{\mathcal{S},\$}$ *is invertible at infinity if and only if its operator spectrum* $\sigma^{\mathrm{op}}(A)$ *is uniformly invertible.*

The term "uniformly invertible" means that

    ①   all elements A_h of $\sigma^{\mathrm{op}}(A)$ are invertible, and
    ②   their inverses are uniformly bounded, $\sup \|A_h^{-1}\| < \infty$.

This theorem yields the vertical arrow in our picture on page ix, and, in a sense, it is the heart of the whole theory and the justification of the study of limit operators.

A big question, that is as old as the first versions of Theorem 1 itself, is whether or not condition ② is redundant. On the one hand, the presence of condition ② often makes the application of Theorem 1 technically difficult. In his review of the article [61], ALBRECHT BÖTTCHER justifiably points out that "Condition ② is nasty to work with." There is nothing to add to this. On the other hand, we do not know of a single example where ② is not redundant, which is why we ask this question. We will address this issue, and we will single out at least some subclasses of $\mathrm{BDO}^p_{\mathcal{S},\$}$ for which the "nasty condition" is indeed known to be redundant.

Equipped with this tool, the limit operator concept, we can now study Fredholmness and stability. The following picture should be seen as a rough guide to this book.

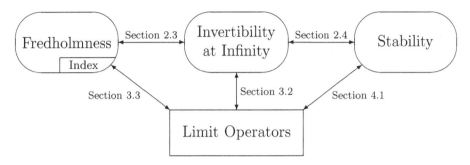

We relate Fredholmness, including the computation of the Fredholm index $\alpha - \beta$, directly to the operator spectrum of A. Moreover, we formulate sufficient and necessary criteria on the applicability of the finite section method (2) in terms of limit operators of the operator A under consideration.

For a brief history of the whole subject, see Sections 1.8, 2.5, 3.10 and 4.4, at the end of each chapter.

About this Book

The original intention of this book was to enrich my PhD thesis by a number of remarks, examples and explanations to increase its readability and to make it accessible to a larger audience. During the actual process of writing this book I found myself slightly deviating from this goal. The present book did not only grow around my thesis, it also became an introductory text to the subject with several branches reaching up to the current frontier of research. It includes a number of new contributions to both theory and applications of band-dominated operators and their limit operators.

There is a noticeable focus on readability in this book. It contains many examples, figures, and remarks, coupled with a healthy amount of very human language. The main ideas and the main actors, *band-dominated operators*, *invertibility at infinity* and *limit operators*, are introduced and illustrated by looking at them from different angles, which might be helpful for readers with various backgrounds.

There is naturally a lot of overlap with the book [70], "Limit Operators and their Applications in Operator Theory" by VLADIMIR RABINOVICH, STEFFEN ROCH and BERND SILBERMANN, published at Birkhäuser in 2004, which is and will be the 'bible' of the limit operator business. However, the non-specialist might appreciate the introductory character and the considerable effort expended on the presentation and readability of the present book. Moreover, it should be mentioned that this book covers a number of topics not included in [70]. Most notably, and this was the main thrust of my PhD thesis, this book treats the spaces ℓ^p with $1 \leq p \leq \infty$ rather than just $1 < p < \infty$. The case $p = 1$ is interesting in, for example, stochastic theory, while the treatment of $p = \infty$ opens the door to the study of operators on the space BC of bounded and continuous functions, to mention only one example. We demonstrate the latter in Sections 4.2.3 and 4.3, and we discuss an application of the developed techniques to boundary integral equations on unbounded rough surfaces.

I experienced the work on this book as equally breathtaking and delightful, and I would be very pleased if the reader will occasionally sense that, too.

Marko Lindner
Reading in May 2006

Acknowledgements

First of all, I am grateful to my friends and colleagues BERND SILBERMANN, STEFFEN ROCH and SIMON CHANDLER-WILDE for their support, for countless fruitful conversations, inspirations and for sharing their expertise with me. I would also like to thank VLADIMIR RABINOVICH for bringing up this beautiful subject and inspiring all of this research in the last decade.

Secondly, I would like to express my gratitude to the above four persons and to ALBRECHT BÖTTCHER, DAVID NEEDHAM and STEPHEN LANGDON for reading parts of the manuscript and making very helpful and valuable comments.

Moreover, I want to thank WOLFGANG SPRÖSSIG for encouraging me to write this book and THOMAS HEMPFLING for his friendly guidance and expertise during all stages of its implementation.

Parts of this research were funded by a Marie Curie Fellowship of the European Union (MEIF-CT-2005-009758). All opinions expressed in this book are those of the author, and not necessarily those of the Commission.

Big thanks go to my friends and colleagues SUE, STEVE, SIMON and DAVE here at Reading, for sharing their time, laughter and wisdom with me and for making me feel comfortable from the very first day and getting me in the right mood to finish this project.

Finally, and on top of all, I am thankful to someone very special for being my muse, holding my hand and believing in me all the time:

Danke, Diana!

Contents

Chapter 1

Preliminaries

1.1 Basic Conventions

1.1.1 Numbers and Vectors

As usual, by $\mathbb{N}$, $\mathbb{Z}$, $\mathbb{Q}$, $\mathbb{R}$ and $\mathbb{C}$ we denote the sets of natural, integer, rational, real and complex numbers, respectively. The positive half axis $(0, +\infty)$ will be abbreviated by $\mathbb{R}_+$, the set of nonnegative integers $\{0, 1, \ldots\}$ is $\mathbb{N}_0$, and the unit circle in the complex plane; that is $\{z \in \mathbb{C} : |z| = 1\}$, is denoted by $\mathbb{T}$.

Throughout the following, n is some natural number used as dimension in $\mathbb{Z}^n$ and $\mathbb{R}^n$. The symbol H will be used as abbreviation of the hypercube $[0, 1)^n$. In the following, we will mostly omit the prefix "hyper-" in "hypercube".

If $U \subset \mathbb{R}^n$ is measurable, then we denote its Lebesgue measure by $|U|$, and for a set $U \subset \mathbb{Z}^n$, its counting measure is denoted by $\#U$.

For every vector $x = (x_1, \ldots, x_n) \in \mathbb{R}^n$, we put $|x| := \max(|x_1|, \ldots, |x_n|)$, and for two sets $U, V \subset \mathbb{R}^n$, we define their distance by

$$\mathrm{dist}\,(U, V) := \inf_{u \in U,\, v \in V} |u - v|.$$

For a real number $x \in \mathbb{R}$, denote its integer part by $[x] := \max\{z \in \mathbb{Z} : z \leq x\}$. Without introducing a new notation, put $[x] := ([x_1], \ldots, [x_n])$ for a vector $x = (x_1, \ldots, x_n) \in \mathbb{R}^n$, so that $x - [x] \in H$ for all $x \in \mathbb{R}^n$.

Remark 1.1. Note that the decision for the maximum norm in $\mathbb{R}^n$ implies that $|x|$ and $\mathrm{dist}\,(U, V)$ are integer if $x \in \mathbb{Z}^n$ and $U, V \subset \mathbb{Z}^n$, which we will find very convenient for the study of band operators, for example. Moreover, balls $\{x \in \mathbb{R}^n : |x| \leq r\}$ in this norm are just cubes $[-r, r]^n$, which will simplify our notations at several points.

However, since in $\mathbb{R}^n$ all norms are equivalent, all of the following theory, apart from a slight modification of what the band-width of a band operator is, also holds if we replace the maximum norm by any other norm in $\mathbb{R}^n$. $\quad\square$

1.1.2   Banach Spaces and Banach Algebras

If not stated otherwise, the letter $\mathbf{X}$ always stands for some complex Banach space; that is a normed vector space over the complex numbers which is complete in its norm. For brevity, we will henceforth refer to this as a *Banach space*.

When talking about a *Banach algebra*, we always mean a unital complex Banach algebra; that is a Banach space $\mathbf{B}$ with another binary operation $\cdot$ which is associative, bilinear and compatible with the norm in $\mathbf{B}$ in the sense that

$$\|x \cdot y\| \ \le \ \|x\| \, \|y\| \qquad \text{for all} \qquad x, y \in \mathbf{B},$$

where in addition, we suppose that there is a *unit element* e in $\mathbf{B}$ such that $e \cdot x = x = x \cdot e$ for all $x \in \mathbf{B}$. Note that in this case, the norm in $\mathbf{B}$ can always be chosen such that $\|e\| = 1$, which is what we will suppose from this point.

As usual, we abbreviate $x \cdot y$ by xy, and we say that $x \in \mathbf{B}$ is *invertible* in $\mathbf{B}$ if there exists an element $y =: x^{-1} \in \mathbf{B}$ such that $xy = e = yx$.

Moreover, when talking about an *ideal* in a Banach algebra $\mathbf{B}$ we always have in mind a closed two-sided ideal; that is a Banach subspace $\mathbf{J}$ of $\mathbf{B}$ such that $bj \in \mathbf{J}$ and $jb \in \mathbf{J}$ whenever $b \in \mathbf{B}$ and $j \in \mathbf{J}$.

If $\mathbf{B}$ is a Banach algebra and M is a subset of $\mathbf{B}$, then $\mathrm{alg}_{\mathbf{B}} M$, $\mathrm{closalg}_{\mathbf{B}} M :=$ $\mathrm{clos}_{\mathbf{B}} (\mathrm{alg}_{\mathbf{B}} M)$ and $\mathrm{closid}_{\mathbf{B}} M$ denote the smallest subalgebra, the smallest Banach subalgebra and the smallest ideal of $\mathbf{B}$ containing M, respectively.

As usual, for an ideal $\mathbf{J}$ in a Banach algebra $\mathbf{B}$, the set

$$\mathbf{B/J} \ := \ \{b + \mathbf{J} \ : \ b \in \mathbf{B}\}$$

with operations

$$(a + \mathbf{J}) + (b + \mathbf{J}) := (a + b) + \mathbf{J}, \qquad \|b + \mathbf{J}\| := \inf_{j \in \mathbf{J}} \|b + j\|, \qquad a, b \in \mathbf{B}$$

is a Banach algebra again, referred to a the *factor algebra of $\mathbf{B}$ modulo $\mathbf{J}$*.

A Banach subalgebra $\mathbf{B}$ of a Banach algebra $\mathbf{A}$ is called *inverse closed* in $\mathbf{A}$ if, whenever $x \in \mathbf{B}$ is invertible in $\mathbf{A}$, also its inverse x^{-1} is in $\mathbf{B}$.

If $\mathbf{B}$ is a Banach algebra and $x \in \mathbf{B}$, then the set

$$\mathrm{sp}_{\mathbf{B}} x \ := \ \{\lambda \in \mathbb{C} \ : \ x - \lambda e \ \text{is not invertible in } \mathbf{B}\}$$

is the *spectrum of x in $\mathbf{B}$*. Spectra are always non-empty compact subsets of $\mathbb{C}$.

Here are two basic results from the theory of Banach algebras of which we will make several uses in what follows. For completeness, we include the short proofs.

Lemma 1.2. *Let $\mathbf{B}$ be a Banach algebra with unit e.*

a) *If $x \in \mathbf{B}$ and $\|x\| < 1$, then $e - x$ is invertible in $\mathbf{B}$, and $\|(e - x)^{-1}\| \le \frac{1}{1 - \|x\|}$.*

b) *If $a \in \mathbf{B}$ is invertible and $x \in \mathbf{B}$ with $\|x\| < \frac{1}{2\|a^{-1}\|}$, then $a - x$ is invertible in $\mathbf{B}$, and $\|(a - x)^{-1}\| \le 2\|a^{-1}\|$.*

Proof. a) One easily checks that the so-called Neumann series $y := e + x + x^2 + \cdots$ converges (even absolutely), and y is the inverse of $e - x$. From the geometric series formula, we get $\|y\| \leq \|x\|^0 + \|x\|^1 + \|x\|^2 + \cdots = \frac{1}{1 - \|x\|}$.

   b) This is a simple consequence of a) and $a - x = a(e - a^{-1}x)$. $\square$

Lemma 1.3. *Consider a Banach algebra* **B** *with unit* e *and a convergent sequence* $(x_m)_{m=1}^\infty$ *of invertible elements in* **B**. *If* $\|x_m^{-1}\| < M < \infty \ \forall m \in \mathbb{N}$ *for some* $M > 0$, *then also the limit* $x := \lim\limits_{m \to \infty} x_m$ *in* **B** *is invertible, and* $\|x^{-1}\| \leq M$.

Proof. From

$$\|x_k^{-1} - x_m^{-1}\| = \|x_k^{-1}(x_m - x_k)x_m^{-1}\| < M^2\|x_m - x_k\|$$

we conclude that also $(x_m^{-1})_{m=1}^\infty$ is a Cauchy sequence in **B**. Denote its limit by y; then we have $x_m \to x$ and $x_m^{-1} \to y$ as $m \to \infty$, and consequently $e = x_m x_m^{-1} \to xy$ and $e = x_m^{-1} x_m \to yx$ which shows that $y = x^{-1}$. Clearly, $\|y\| \leq M$. $\square$

1.1.3 Operators

By $L(\mathbf{X})$ we denote the set of all bounded and linear operators A on the Banach space **X** which, equipped with point-wise addition and scalar multiplication and the usual operator norm

$$\|A\| := \sup_{x \neq 0} \frac{\|Ax\|_\mathbf{X}}{\|x\|_\mathbf{X}} = \sup_{x \in \mathbf{X}, \|x\|_\mathbf{X} = 1} \|Ax\|_\mathbf{X},$$

is a Banach space as well. Using the composition of two operators as multiplication in $L(\mathbf{X})$, it is also a Banach algebra with unit I, the identity operator on **X**.

   As usual, we say that an operator $A \in L(\mathbf{X})$ is *invertible*, if it is an invertible element of the Banach algebra $L(X)$. This is the case if and only if $A : \mathbf{X} \to \mathbf{X}$ is bijective, since, by a theorem of BANACH, this already implies the linearity and boundedness of the inverse operator A^{-1}.

   Let $A, A_1, A_2, \ldots \in L(\mathbf{X})$ be arbitrary operators. We will say that the sequence $A_1, A_2, \ldots$ *converges strongly* to A as $m \to \infty$, and write $A_m \to A$, if

$$\|A_m x - Ax\|_\mathbf{X} \to 0 \quad \text{as} \quad m \to \infty \quad \text{for every} \quad x \in \mathbf{X}.$$

On the other hand, we will say that the sequence $A_1, A_2, \ldots$ *norm-converges* to A as $m \to \infty$, and write $A_m \rightrightarrows A$, if

$$\|A_m - A\|_{L(\mathbf{X})} \to 0 \quad \text{as} \quad m \to \infty.$$

Further, let $[A, B]$ refer to the *commutator* of two operators $A, B \in L(\mathbf{X})$; that is

$$[A, B] = AB - BA.$$

As usual, let $\ker A = \{x \in \mathbf{X} : Ax = 0\}$ and $\operatorname{im} A = \{Ax : x \in \mathbf{X}\}$ denote the *kernel* and the *image* (or *range*) of the operator $A \in L(\mathbf{X})$.

1.2   The Spaces

We will study both spaces of functions $\mathbb{R}^n \to \mathbf{X}$ and spaces of functions $\mathbb{Z}^n \to \mathbf{X}$; the latter functions will henceforth be referred to as (multi-dimensional) sequences. Another terminology for these two cases that will often be used is "the continuous case" and "the discrete case". Note that the words "continuous" and "discrete" describe the domains $\mathbb{R}^n$ and $\mathbb{Z}^n$, respectively – not a property of the functions.

1.2.1   Functions

Let T be some nonempty measurable subset of $\mathbb{R}^n$, and $\mathbf{X}$ be a Banach space. Then, for all $p \in [1, \infty)$, we refer to the set of all (equivalence classes of) Lebesgue measurable functions $f : T \to \mathbf{X}$, for which the p-th power of $\|f(.)\|_{\mathbf{X}}$ is Lebesgue integrable, by $L^p(T, \mathbf{X})$. By $L^\infty(T, \mathbf{X})$ we denote the set of all (equivalence classes of) Lebesgue measurable functions $f : T \to \mathbf{X}$ that are essentially bounded. The $\mathbf{X}$ is omitted in these notations if $\mathbf{X} = \mathbb{C}$. Moreover, if $n \in \mathbb{N}$ is fixed, then $L^p(\mathbb{R}^n)$ will be abbreviated by L^p for all $p \in [1, \infty]$.

Equipped with point-wise addition and scalar multiplication as well as the usual Lebesgue integral norm

$$\|f\|_p = \sqrt[p]{\int_T \|f(t)\|_{\mathbf{X}}^p \, dt} \qquad \text{for} \qquad p < \infty$$

and the essential supremum norm

$$\|f\|_\infty = \operatorname*{ess\,sup}_{t \in T} \|f(t)\|_{\mathbf{X}}$$
$$= \inf \left\{ M > 0 : |\{t \in T : \|f(t)\|_{\mathbf{X}} > M\}| = 0 \right\} \qquad \text{for} \qquad p = \infty,$$

the sets $L^p(T, \mathbf{X})$, $p \in [1, \infty]$ become Banach spaces. If $\mathbf{X}$ is a Banach algebra, then, in addition, $L^\infty(T, \mathbf{X})$ can also be equipped with a point-wise multiplication which makes this space a Banach algebra as well.

1.2.2   Sequences

Analogously to the function spaces, the usual Banach spaces $\ell^p(S, \mathbf{X})$ and $\ell^\infty(S, \mathbf{X})$ of $\mathbf{X}$-valued sequences on $S \subset \mathbb{Z}^n$ are defined. So for $u = (u_\alpha)_{\alpha \in S}$, we put

$$\|u\|_p := \sqrt[p]{\sum_{\alpha \in S} \|u_\alpha\|_{\mathbf{X}}^p} \qquad \text{for} \qquad p < \infty,$$
$$\|u\|_\infty := \sup_{\alpha \in S} \|u_\alpha\|_{\mathbf{X}} \qquad \text{for} \qquad p = \infty.$$

Again, $\mathbf{X}$ is omitted in these notations if $\mathbf{X} = \mathbb{C}$, and moreover $\ell^p(\mathbb{Z}^n)$ is abbreviated by ℓ^p for $p \in [1, \infty]$, if $n \in \mathbb{N}$ is fixed. Note that, again, $\ell^\infty(S, \mathbf{X})$ is a Banach algebra with componentwise defined operations if $\mathbf{X}$ is one.

1.2.3 Discretization: Functions as Sequences

Often it is sufficient to develop a theory for sequences only and to regard functions on $\mathbb{R}^n$ as sequences with some slightly more sophisticated values. This can be done by a rather simple and natural construction: We cut $\mathbb{R}^n$ into cubes and identify a function on $\mathbb{R}^n$ with the sequence of its restrictions to these cubes.

Therefore, remember that H denotes the cube $[0,1)^n$, and define $H_\alpha := \alpha + H$ for $\alpha \in \mathbb{Z}^n$. Clearly, $\{H_\alpha\}_{\alpha \in \mathbb{Z}_n}$ is a disjoint decomposition of $\mathbb{R}^n$. Let $p \in [1, \infty]$, and let G denote the operator that assigns to every function $f \in L^p$ the sequence of its restrictions to the cubes H_α,

$$G : f \mapsto (f|_{H_\alpha})_{\alpha \in \mathbb{Z}^n}. \tag{1.1}$$

By naturally identifying $L^p(H_\alpha)$ with $L^p(H)$ (via shift by α), we get that G is an isometrical isomorphism from L^p to the sequence space $\ell^p(\mathbb{Z}^n, L^p(H))$ (in the notation of Section 1.2.2 with $L^p(H)$ as the Banach space $\mathbf{X}$). With this isometrical isomorphism in mind, we henceforth write

$$L^p \cong \ell^p\Big(\mathbb{Z}^n, L^p(H)\Big). \tag{1.2}$$

In the same way we can handle the more general case $L^p(\mathbb{R}^n, \mathbf{X})$, for which we have that $L^p(\mathbb{R}^n, \mathbf{X}) \cong \ell^p(\mathbb{Z}^n, L^p(H, \mathbf{X}))$, by means of G.

As a consequence, every operator A on L^p can be identified with the operator

$$A^G := GAG^{-1} \tag{1.3}$$

on $\ell^p(\mathbb{Z}^n, L^p(H))$, which will be referred to as the *discretization* of A.

1.2.4 The System Case

For simplicity of notation (and imagination), in the continuous case, we will mainly restrict ourselves to spaces of scalar-valued functions and to operators on such. But it is worth mentioning that most of our investigations and results remain valid if we pass to spaces of vector-valued functions.

For example, if n and N are integers, $p \in [1, \infty]$ and $T \subset \mathbb{R}^n$, then a little thought shows that a function f is in $L^p(T, \mathbb{C}^N)$ if and only if it $f = (f_1, \ldots, f_N)$ with scalar-valued functions $f_i \in L^p(T, \mathbb{C}) = L^p(T)$. Consequently, we can think of every operator A on $L^p(T, \mathbb{C}^N)$ as a $N \times N$-matrix whose entries A_{ij} are operators on $L^p(T)$. This is what usually is called "the system case".

1.3 The Operators

1.3.1 Operators of Shift and Multiplication

Fix some Banach space $\mathbf{X}$. First we introduce two very simple classes of operators on sequences, which will turn out to be the basic building stones for the operators to be studied in this book.

Definition 1.4. *If $b = (b_\alpha)_{\alpha \in \mathbb{Z}^n}$ is a bounded sequence of operators $b_\alpha \in L(\mathbf{X})$, then by $\hat{M}_b$ we will denote the generalized multiplication operator, acting on every $u \in \ell^p(\mathbb{Z}^n, \mathbf{X})$ by*

$$(\hat{M}_b u)_\alpha = b_\alpha u_\alpha \qquad \forall \alpha \in \mathbb{Z}^n.$$

Often we will refer to b as the symbol of $\hat{M}_b$.

Definition 1.5. *For every $\alpha \in \mathbb{Z}^n$, we denote the shift operator on $\ell^p(\mathbb{Z}^n, \mathbf{X})$ by V_α, acting by*

$$(V_\alpha u)_\beta = u_{\beta - \alpha} \qquad \forall \beta \in \mathbb{Z}^n,$$

i.e. shifting the whole sequence $u \in \ell^p(\mathbb{Z}^n, \mathbf{X})$ by the vector $\alpha \in \mathbb{Z}^n$.

As already mentioned, it is often convenient to treat L^p as a certain ℓ^p-space (see (1.2)). But having a look at one of the aims of this book – approximation methods in L^p – it becomes clear that L^p has certainly more interest for us than as just a nice and illustrative example of a space $\ell^p(\mathbb{Z}^n, \mathbf{X})$ with $\dim \mathbf{X} = \infty$ and thus sometimes needs special treatment. Therefore, we will adapt Definitions 1.4 and 1.5 and give analogous but more appropriate definitions for L^p.

Definition 1.6. *For every function $b \in L^\infty$, let $M_b \in L(L^p)$ denote the operator of multiplication (or multiplicator) by b, acting on every $u \in L^p$ by*

$$(M_b u)(x) = b(x)u(x) \qquad \forall x \in \mathbb{R}^n.$$

Frequently, we will call b the symbol of the multiplicator M_b.

Definition 1.7. *Without introducing a new symbol, let V_κ denote the shift operator on L^p by $\kappa \in \mathbb{R}^n$, acting by the rule*

$$(V_\kappa u)(x) = u(x - \kappa) \qquad \forall x \in \mathbb{R}^n,$$

i.e. shifting the whole function $u \in L^p$ by the vector $\kappa \in \mathbb{R}^n$.

Obviously, the discretization (1.3) of M_b is a generalized multiplication operator, and the discretization of V_κ with $\kappa = \alpha \in \mathbb{Z}^n$ is the (discrete) shift V_α.

1.3.2 Adjoint and Pre-adjoint Operators

As usual, the *dual space* of a Banach space $\mathbf{X}$; that is the space of all bounded and linear functionals on $\mathbf{X}$, is denoted by $\mathbf{X}^*$. Moreover, if it exists and is unique, then by $\mathbf{X}^\triangleleft$ we denote the Banach space whose dual space is (isometrically isomorphic to) $\mathbf{X}$, and we will refer to $\mathbf{X}^\triangleleft$ as the *pre-dual space* of $\mathbf{X}$.

Remark 1.8. Note that in general, neither existence nor uniqueness (up to isometrical isomorphy, of course) of a pre-dual space is guaranteed. A simple counter-example is ℓ^1 which has the two different pre-dual spaces c and c_0, the spaces of all convergent and all null sequences, respectively.

However, it was first pointed out by GROTHENDIECK [33] that all ℓ^∞- and L^∞-spaces have a unique pre-dual, and this observation was generalized to von Neumann algebras later in [77]. $\qquad\qquad\square$

Remember that $\mathbf{X}$ can be identified with a subspace $\hat{\mathbf{X}}$ of $\mathbf{X}^{**}$ by the mapping

$$x \in \mathbf{X} \; \mapsto \; \hat{x} \in \hat{\mathbf{X}} \quad \text{with} \quad \hat{x}(f) = f(x) \quad \forall f \in \mathbf{X}^*.$$

As usual, we call $\mathbf{X}$ *reflexive*, if $\hat{\mathbf{X}} = \mathbf{X}^{**}$. Moreover, recall that for $A \in L(\mathbf{X})$, the *adjoint operator* $A^* \in L(\mathbf{X}^*)$ is defined by

$$(A^* f)(x) = f(Ax) \quad \text{for all} \quad x \in \mathbf{X} \quad \text{and} \quad f \in \mathbf{X}^*.$$

It is well-known that, for $p \in [1, \infty)$, the dual space of $\ell^p(\mathbb{Z}^n, \mathbf{X})$ can be identified with $\ell^q(\mathbb{Z}^n, \mathbf{X}^*)$, where $1/p + 1/q = 1$ and $1/\infty = 0$. Unfortunately, this is not true for $p = \infty$. The dual space of $\ell^\infty(\mathbb{Z}^n, \mathbf{X})$ is not isomorphic to $\ell^1(\mathbb{Z}^n, \mathbf{X}^*)$ – it is strictly larger[1]. That makes the identification and study of the adjoint operator A^* of $A \in L(\ell^p(\mathbb{Z}^n, \mathbf{X}))$ much more difficult for $p = \infty$ than for $p < \infty$.

For some arguments in the case $p = \infty$, where the aspect of duality is important, we will therefore need to find an adequate substitute for the adjoint operator A^*. Fix a Banach space E, and by F denote the pre-dual space of E. In our case, $E = \ell^\infty(\mathbb{Z}^n, \mathbf{X})$, the space F is isometrically isomorphic to $\ell^1(\mathbb{Z}^n, \mathbf{X}^\triangleleft)$ – provided that $\mathbf{X}^\triangleleft$ exists (see e.g. [77]).

If $A \in L(E)$, $F \cong E^\triangleleft$ and if there exists an operator $B \in L(F)$ such that

$$B^* = A, \tag{1.4}$$

we will refer to B as the *pre-adjoint operator* of A, the operator whose adjoint equals A, and we will frequently denote B by $A^\triangleleft$. In many situations we will restrict ourselves to operators A on E that possess a pre-adjoint operator.

There is an alternative and equivalent characterization of those operators $A \in L(E)$ that possess a pre-adjoint operator. Again suppose that the pre-dual space F of E exists, and remember that F can be identified with $\hat{F} \subset F^{**} \cong E^*$.

If the adjoint operator A^*, acting on E^*, maps the subspace $\hat{F} \subset E^*$ into $\hat{F}$ again, we can define an operator $B \in L(F)$ by

$$\widehat{Bf} = A^* \hat{f} \quad \forall f \in F. \tag{1.5}$$

Proposition 1.9. *If F is a Banach space, $E = F^*$, and $A \in L(E)$, then $B \in L(F)$ is the pre-adjoint of A, i.e. (1.4) holds if and only if $A^*(\hat{F}) \subset \hat{F}$ and (1.5) holds.*

Proof. For arbitrary elements $e \in E$ and $f \in F$ and arbitrary operators $A \in L(E)$ and $B \in L(F)$, one has

$$(Ae)(f) = \hat{f}(Ae) = (A^* \hat{f})(e)$$

and

$$(B^* e)(f) = e(Bf) = (\widehat{Bf})(e).$$

Consequently, (1.4) implies (1.5), and vice versa. $\qquad\square$

---

[1] See Example 1.26 c) for a functional on $\ell^\infty(\mathbb{Z}^n, \mathbf{X})$ which does not correspond to an element of $\ell^1(\mathbb{Z}^n, \mathbf{X}^*)$.

Now in our case, $E = \ell^\infty(\mathbb{Z}^n, \mathbf{X})$ and $F = \ell^1(\mathbb{Z}^n, \mathbf{X}^\triangleleft)$, we denote the set of all $A \in L(E)$ that possess a pre-adjoint operator $A^\triangleleft \in L(F)$ by

$$\mathcal{S} := \left\{ A = B^* \in L(E) \; : \; B \in L(F) \right\} = \left\{ A \in L(E) \; : \; A^*(\hat{F}) \subset \hat{F} \right\}.$$

So for $A \in \mathcal{S}$, we can pass from $L(E)$ to $L(F)$ by $A \mapsto A^\triangleleft$ and back to $L(E)$ by $B \mapsto B^*$. From basic properties of the adjoint operator it follows that $A^\triangleleft$ is invertible in $L(F)$ if and only if $(A^\triangleleft)^* = A$ is invertible in $L(E)$. Moreover,

$$\|A^\triangleleft\|_{L(F)} = \|(A^\triangleleft)^*\|_{L(F^*)} = \|A\|_{L(E)}. \tag{1.6}$$

Proposition 1.10. *$\mathcal{S}$ is an inverse closed Banach subalgebra of $L(E)$.*

Proof. If $A_1 = B_1^*$ and $A_2 = B_2^*$ are in $\mathcal{S}$, then also $A_1 + A_2 = (B_1 + B_2)^*$ and $A_1 A_2 = (B_2 B_1)^*$ are in $\mathcal{S}$. If $(A_k) \subset \mathcal{S}$ tends to A in the norm of $L(E)$, then, by (1.6), not only (A_k) is a Cauchy sequence in $L(E)$ but also the sequence $(B_k) = (A_k^\triangleleft)$ is a Cauchy sequence in $L(F)$. Let B denote the norm limit of B_k. Then, by (1.6) again, $A_k = B_k^* \rightrightarrows B^*$. Together with $A_k \rightrightarrows A$, this shows that $A = B^* \in \mathcal{S}$.

To see that $\mathcal{S}$ is inverse closed, take an arbitrary invertible operator $A \in \mathcal{S}$. But then, also $B = A^\triangleleft$ is invertible, and $A^{-1} = (B^*)^{-1} = (B^{-1})^* \in \mathcal{S}$. $\qquad\square$

1.3.3   An Approximate Identity

Primarily – but not only – in connection with approximation methods, we will deal with the following projection operators.

Definition 1.11. *Consider a set $U \subset \mathbb{Z}^n$. We define P_U, acting on $u \in \ell^p(\mathbb{Z}^n, \mathbf{X})$ by*

$$(P_U u)_\alpha = \begin{cases} u_\alpha & \text{if } \alpha \in U, \\ 0 & \text{if } \alpha \notin U. \end{cases}$$

Clearly, P_U is a projector. We will refer to its complementary projector $I - P_U$ by Q_U. Typical examples of projectors P_U we have to deal with are of the form $P_k := P_{\{-k,\dots,k\}^n}$ with some $k \in \mathbb{N}$. Moreover, put $Q_k := I - P_k$.

Definition 1.12. *For every measurable set $U \subset \mathbb{R}^n$, put $P_U := M_{\chi_U} \in L(L^p)$, where χ_U is the characteristic function of U. Again, P_U and $Q_U := I - P_U$ are complementary projectors. Also here we spend some extra notation on the typical examples: For $\tau \in \mathbb{R}_+$, put $P_\tau := P_{[-\tau,\tau]^n}$ and $Q_\tau := I - P_\tau$.*

In connection with approximation methods, we need a sequence of such projectors, that is increasing in an appropriate sense. We will use the sequence

$$\mathcal{P} := (P_1, P_2, P_3, \dots)$$

where, depending on whether we are in the discrete or in the continuous case, $P_1, P_2, P_3, \dots$ are those from Definition 1.11 or 1.12, respectively. Note that, in the continuous case L^p, for very $k \in \mathbb{N}$, the discretization P_k^G is exactly the operator P_k on the discrete space $\ell^p(\mathbb{Z}^n, L^p(H))$ associated with L^p by (1.2).

In either case, $\mathcal{P}$ is an *approximate identity* in the terminology of [70]; precisely, it is subject to the constraints (that are a bit stronger than those in [70])

$$P_k P_m \;=\; P_{\min(k,m)}$$

$$\text{and} \quad \|P_m u\| \;\to\; \|u\| \quad \text{as} \quad m \to \infty \tag{1.7}$$

for every element u of the space under consideration. Our approximate identity $\mathcal{P}$ is the natural candidate for the construction of approximation methods like (2). The projection operators P_m are also referred to as *finite section projectors*.

1.3.4   Compact Operators and their Substitutes

Fix some $p \in [1, \infty]$ and a Banach space $\mathbf{X}$. In what follows, E stands for the space $\ell^p(\mathbb{Z}^n, \mathbf{X})$, and $\mathcal{P} = (P_1, P_2, \ldots)$ is the corresponding approximate identity.

Compact Operators and Strong Convergence

By $K(E) \subset L(E)$ we denote the ideal of compact operators on E. One crucial property of operators $T \in K(E)$ is that they turn strong convergence into norm convergence if they are applied to the convergent sequence from the right. That is, $A_m \to A$ implies

$$A_m T \rightrightarrows AT \quad \text{as} \quad m \to \infty. \tag{1.8}$$

This implication also holds the other way round (Theorem 1.1.3 in [70]):

Proposition 1.13. $A_m \to A$ on E if and only if (1.8) holds for all $T \in K(E)$.

Sometimes we are also interested in sequences (A_m) for which the symmetric counter-part of (1.8) is true; that is,

$$TA_m \rightrightarrows TA \quad \text{as} \quad m \to \infty \tag{1.9}$$

for all $T \in K(E)$. This property is closely related with the strong convergence of the adjoints as well as pre-adjoints of A_m.

Proposition 1.14. *If $A_m^* \to A^*$ strongly on E^*, then (1.9) holds for all $T \in K(E)$, which moreover implies the strong convergence $A_m^\triangleleft \to A^\triangleleft$ on $E^\triangleleft$, provided that $E^\triangleleft$ and the pre-adjoint operators exist.*

Proof. If $A_m^* \to A^*$ on E^*, then, for all $T_1 \in K(E^*)$, $\|(A_m^* - A^*)T_1\| \to 0$ by Proposition 1.13. By Schauder's theorem [88, p. 282], $\{T^* : T \in K(E)\}$ is a subset of $K(E^*)$ – but it is a proper[2] subset if E is irreflexive. Consequently, for all $T \in K(E)$,

$$\|T(A_m - A)\| = \|(T(A_m - A))^*\| = \|(A_m^* - A^*)T^*\| \to 0 \quad \text{as} \quad m \to \infty.$$

---

[2]For an operator in the difference set, see Example 1.26 c) with $E = \ell^1(\mathbb{Z}, \mathbf{X})$.

If $E^\lhd$ and the pre-adjoint operators of A_m and A exist, and (1.9) holds for all $T \in K(E)$, then

$$\|(A_m^\lhd - A^\lhd)T_2\| = \|((A_m^\lhd - A^\lhd)T_2)^*\| = \|T_2^*(A_m - A)\| \to 0 \qquad \text{as} \qquad m \to \infty$$

for all $T_2 \in K(E^\lhd)$, which proves $A_m^\lhd \to A^\lhd$ on $E^\lhd$, by Proposition 1.13 again.   □

Definition 1.15. *We will say that A_m converges $*$-strongly to A if $A_m \to A$ and $A_m^* \to A^*$, that A_m converges pre$*$-strongly to A if $A_m \to A$ and $A_m^\lhd \to A^\lhd$ as $m \to \infty$, and that A_m converges K-strongly to A if (1.8) and (1.9) hold for all $T \in K(E)$. We will write $A_m \overset{*}{\to} A$, $A_m \overset{\lhd}{\to} A$ and $A_m \overset{K}{\to} A$, respectively.*

Corollary 1.16.   a) $A_m \overset{*}{\to} A$ $\implies$ $A_m \overset{K}{\to} A$ $\implies$ $A_m \overset{\lhd}{\to} A$.

b) *If E is reflexive; that is if $\mathbf{X}$ is reflexive and $1 < p < \infty$, then all three convergence types coincide.*

Proof. a) immediately follows from Propositions 1.13 and 1.14.

b) Clearly, if E is reflexive, then $E^\lhd \cong E^*$, and all pre-adjoint operators coincide with the adjoint operators, whence the claim directly follows from a).   □

Proposition 1.13 and Corollary 1.16 b) specify what we mean by saying that the set $K(E)$ of compact operators on E **determines** strong convergence and, if E is reflexive, also $*$-strong convergence of operators on E, in terms of (1.8) (and (1.9)).

Substituting $K(E)$

The name "approximate identity" indicates that the sequence $\mathcal{P} = (P_1, P_2, \ldots)$, in some sense, approximates the identity operator I on E. But in general this is neither of the convergence types in Definition 1.15. Although we have $P_m \overset{*}{\to} I$ for $1 < p < \infty$, this is not true for $p \in \{1, \infty\}$. The key problem is that the sequence $\mathcal{P} = (P_m)$ is not strongly convergent to I if $p = \infty$.

In terms of compact operators this means that

$$P_m T \rightrightarrows IT, \qquad \text{i.e.} \qquad Q_m T \rightrightarrows 0 \qquad \text{as} \qquad m \to \infty \qquad (1.10)$$

holds for all $T \in K(E)$, if and only if $p < \infty$. Even worse, the symmetric property

$$T P_m \rightrightarrows TI, \qquad \text{i.e.} \qquad T Q_m \rightrightarrows 0 \qquad \text{as} \qquad m \to \infty \qquad (1.11)$$

holds for all $T \in K(E)$ if and only[3] if $1 < p < \infty$.

---

[3] Take $T = \tilde{A}$ and $T = C$ from Example 1.26 as counter-example for $p = 1$ and $p = \infty$, respectively.

These unsatisfactory restrictions on p show that, although $K(E)$ determines strong and, in the reflexive case, even $*$-strong convergence, it is in general not the appropriate class of operators to determine the way in which $\mathcal{P}$ approximates the identity I. Our consequence is to substitute $K(E)$ by this more appropriate class of operators:

Definition 1.17. *By $K(E, \mathcal{P})$ we denote the set of all operators $T \in L(E)$ for which (1.10) and (1.11) are true.*

Unfortunately, unlike $K(E)$, its substitute $K(E, \mathcal{P})$ is no longer an ideal in $L(E)$. That is why we introduce the largest subalgebra of $L(E)$ that $K(E, \mathcal{P})$ is an ideal of:

Definition 1.18. *By $L(E, \mathcal{P})$ we denote the set of all operators $A \in L(E)$ for which AT and TA are in $K(E, \mathcal{P})$ whenever $T \in K(E, \mathcal{P})$.*

Proposition 1.19. *$L(E, \mathcal{P})$ is a Banach subalgebra of $L(E)$, and $K(E, \mathcal{P})$ is an ideal in $L(E, \mathcal{P})$.*

Proof. It is obvious that with $\lambda \in \mathbb{C}$ and $T_1, T_2 \in K(E, \mathcal{P})$ also λT_1, $T_1 + T_2$ and $T_1 T_2$ are in $K(E, \mathcal{P})$. To see that $K(E, \mathcal{P})$ is closed, take a sequence $(T_k)_{k=1}^{\infty} \subset K(E, \mathcal{P})$ with $T_k \rightrightarrows T \in L(E)$. Given an $\varepsilon > 0$, choose k large enough that $\|T - T_k\| < \varepsilon/2$, and choose m_0 such that $\|T_k Q_m\| < \varepsilon/2$ for all $m > m_0$. Then also

$$\|T Q_m\| \leq \|T_k Q_m\| + \|T - T_k\| \cdot \|Q_m\| < \frac{\varepsilon}{2} + \frac{\varepsilon}{2} \cdot 1 = \varepsilon$$

for all $m > m_0$. Analogously one shows that also $Q_m T \rightrightarrows 0$ as $m \to \infty$, and hence, also $T \in K(E, \mathcal{P})$.

The construction in Definition 1.18 immediately implies that $L(E, \mathcal{P})$ is a Banach algebra (note that $K(E, \mathcal{P})$ is a closed algebra), which contains $K(E, \mathcal{P})$ as an ideal. $\qquad\square$

Although this is trivial, we remark that all elements P_m of $\mathcal{P}$ are clearly contained in $K(E, \mathcal{P})$. For membership in $L(E, \mathcal{P})$ we have a nice criterion which is similar to properties (1.10) and (1.11), defining $K(E, \mathcal{P})$.

Proposition 1.20. *$A \in L(E)$ is contained in $L(E, \mathcal{P})$ if and only if, for every $k \in \mathbb{N}$,*

$$P_k A Q_m \rightrightarrows 0 \qquad and \qquad Q_m A P_k \rightrightarrows 0 \qquad as \qquad m \to \infty. \qquad (1.12)$$

Proof. First suppose $A \in L(E, \mathcal{P})$ and take a $k \in \mathbb{N}$. From $P_k \in K(E, \mathcal{P})$ we get that $P_k A$, $A P_k \in K(E, \mathcal{P})$ which shows (1.12).

Now suppose that (1.12) is true, and take an arbitrary $T \in K(E, \mathcal{P})$. To see that $AT \in K(E, \mathcal{P})$, note that $(AT)Q_m = A(TQ_m) \rightrightarrows 0$ as $m \to \infty$, and that,

$$\|Q_m(AT)\| \leq \|Q_m A P_k\| \cdot \|T\| + \|Q_m A\| \cdot \|Q_k T\|$$

holds for every $k \in \mathbb{N}$, where, by (1.12), the first term tends to zero as $m \to \infty$, and $\|Q_k T\|$ can be made as small as desired by choosing k large enough. By a symmetric argument, one shows that also $TA \in K(E, \mathcal{P})$, and hence, $A \in L(E, \mathcal{P})$. $\qquad\square$

Corollary 1.21. *For $T \in L(E, \mathcal{P})$, either of the two conditions* (1.10) *and* (1.11) *implies the other one, and hence that $T \in K(E, \mathcal{P})$.*

Proof. Suppose $T \in L(E, \mathcal{P})$ is subject to (1.10). Then, for every $k \in \mathbb{N}$,

$$\|TQ_m\| \leq \|P_k TQ_m\| + \|Q_k T\| \cdot \|Q_m\|$$

holds, where, by Proposition 1.20, the first term tends to zero as $m \to \infty$, and $\|Q_k T\|$ can be made as small as desired by (1.10). A symmetric argument shows that also (1.11) implies (1.10), provided $T \in L(E, \mathcal{P})$. □

Remember that, in general, $P_m \nrightarrow I$ and hence $Q_m \nrightarrow 0$ strongly on E, which lead us to the introduction of $K(E, \mathcal{P})$ and $L(E, \mathcal{P})$. Now put

$$E_0 = \{u \in E : Q_m u \to 0 \text{ as } m \to \infty\}. \tag{1.13}$$

It is easy to check that E_0 is a Banach subspace of E.

Proposition 1.22. *Every operator $A \in L(E, \mathcal{P})$ maps E_0 to E_0.*

Proof. Take arbitrary $A \in L(E, \mathcal{P})$ and $u \in E_0$. Then, for all $k, m \in \mathbb{N}$,

$$\|Q_m Au\| \leq \|Q_m AP_k\| \|u\| + \|Q_m A\| \|Q_k u\|$$

holds, which can be made arbitrarily small by choosing m and k sufficiently large. Consequently, $Au \in E_0$. □

If $p = 2$ and $\dim \mathbf{X} < \infty$, an operator $A \in L(E)$ is called a *quasidiagonal operator* with respect to $\mathcal{P}$ (introduced by HALMOS [37]) if $[P_k, A] \rightrightarrows 0$ holds as $k \to \infty$. It is readily checked that this class is contained in $L(E, \mathcal{P})$, even if we generalize that definition to $p \in [1, \infty]$ and to arbitrary Banach spaces $\mathbf{X}$. Indeed, if A is quasidiagonal and $m \geq k$, then

$$P_k AQ_m = P_k A - P_k AP_m = P_k(P_m A - AP_m) \rightrightarrows 0 \qquad \text{as} \qquad m \to \infty.$$

By a symmetric argument we see that A also has the second property in (1.12), and consequently, $A \in L(E, \mathcal{P})$.

We conclude this discussion with another corollary of Proposition 1.20, yielding to an alternative description of the class $L(E, \mathcal{P})$ in terms of the commutator $[P_k, A]$.

Corollary 1.23. *$A \in L(E)$ is contained in $L(E, \mathcal{P})$ if and only if, for every $k \in \mathbb{N}$,*

$$[P_k, A] \in K(E, \mathcal{P}).$$

Proof. Clearly, if $A \in L(E, \mathcal{P})$, then $P_k A, AP_k \in K(E, \mathcal{P})$, whence also the commutator $[P_k, A] = P_k A - AP_k$ is in $K(E, \mathcal{P})$ for every $k \in \mathbb{N}$.

For the reverse direction, note that for every fixed $k \in \mathbb{N}$,

$$P_k AQ_m = [P_k, A]Q_m + AP_k Q_m \rightrightarrows 0 \qquad \text{as} \qquad m \to \infty$$

since $[P_k, A] \in K(E, \mathcal{P})$ and $P_k Q_m = 0$ for all $m \geq k$. Analogously, we prove the second property in (1.12), showing that $A \in L(E, \mathcal{P})$. □

Inclusions

We have substituted $K(E)$ by $K(E, \mathcal{P})$ and $L(E)$ by $L(E, \mathcal{P})$. It is interesting to see how these four sets are related to each other. Obviously,

$$K(E, \mathcal{P}) \subsetneq L(E, \mathcal{P}) \subset L(E).$$

But it is less obvious how $K(E)$ squeezes into that chain of inclusions. The answer is essentially given by the following proposition.

Proposition 1.24. *The inclusion* $L(E, \mathcal{P}) \cap K(E) \subset K(E, \mathcal{P})$ *is true, and equality holds if and only if* $\mathcal{P} \subset K(E)$.

Proof. Let $A \in L(E, \mathcal{P}) \cap K(E)$. If $\|AQ_m\|$ does not tend to zero as $m \to \infty$, then there exists a bounded sequence $(x_m) \subset E$ such that

$$AQ_m x_m \not\to 0 \qquad \text{as} \qquad m \to \infty. \tag{1.14}$$

From $A \in L(E, \mathcal{P})$ we conclude that, for every fixed $k \in \mathbb{N}$,

$$\|P_k A Q_m x_m\| \leq \|P_k A Q_m\| \sup_m \|x_m\| \to 0 \qquad \text{as} \qquad m \to \infty. \tag{1.15}$$

Since the sequence $(Q_m x_m)$ is bounded and A is compact, each subsequence of $(AQ_m x_m)$ has a convergent subsequence, where, by (1.15), the limit of the latter can be nothing but zero. But that shows that the sequence $(AQ_m x_m)$ itself tends to zero, in contradiction to (1.14).

Consequently, $AQ_m \rightrightarrows 0$ as $m \to \infty$. Together with the fact that $A \in L(E, \mathcal{P})$, this shows that also the symmetric property $Q_m A \rightrightarrows 0$ is true by Corollary 1.21. Consequently, $A \in K(E, \mathcal{P})$, which shows that $L(E, \mathcal{P}) \cap K(E) \subset K(E, \mathcal{P})$.

If $P_m \in K(E)$ for every $m \in \mathbb{N}$ and $A \in K(E, \mathcal{P})$, then not only all operators AP_m but also their norm limit A (see (1.11)) is compact.

If some P_m is not in $K(E)$, then this is an example of an operator in $K(E, \mathcal{P})$ which is not in $L(E, \mathcal{P}) \cap K(E)$. $\qquad\square$

Remark 1.25. For $E = \ell^p(\mathbb{Z}^n, \mathbf{X})$, one has

$$L(E, \mathcal{P}) \cap K(E) = K(E, \mathcal{P}) \qquad \Longleftrightarrow \qquad \mathcal{P} \subset K(E) \qquad \Longleftrightarrow \qquad \dim \mathbf{X} < \infty.$$

If $E = \ell^p(\mathbb{Z}^n, L^p(H)) \cong L^p$, then $\mathcal{P} \not\subset K(E)$ since $\dim L^p(H) = \infty$. Hence, we have a proper inclusion $L(E, \mathcal{P}) \cap K(E) \subsetneq K(E, \mathcal{P})$ in this case. $\qquad\square$

Proposition 1.24 essentially yields the nice pictures in Figure 1 on page 15. But before we come to these Venn diagrams, we first give some basic examples of operators which are in $L(E)$ but not in $L(E, \mathcal{P})$. In all these examples, the first condition in (1.12) is violated. Note that some of these operators are compact and some are not. We will include these operators in the Venn diagrams in Figure 1. For simplicity, we restrict ourselves to the case $n = 1$.

Example 1.26. a) Our first example consists of an operator on $\ell^1(\mathbb{Z}, \mathbf{X})$,

$$A : (u_i) \mapsto \left(\ldots, 0, 0, \sum_{i=-\infty}^{\infty} u_i, 0, 0, \ldots \right),$$

where the sum is in the 0-th component.

Moreover, we consider his compact friend $\tilde{A}$ on $\ell^1(\mathbb{Z}, \mathbf{X})$ with

$$\tilde{A} : (u_i) \mapsto \left(\ldots, 0, 0, \sum_{i=-\infty}^{\infty} f(u_i) \, a, 0, 0, \ldots \right)$$

where $f \in \mathbf{X}^*$ and $a \in \mathbf{X}$ are fixed non-zero elements. Note that, unlike A, the operator $\tilde{A}$ is compact, independently of dim $\mathbf{X}$.

b) Our second example is the operator $B : u \mapsto v$ on L^p with $p \in [1, \infty]$, where

$$v(x) = \begin{cases} u(x + k), & x \in \left(1 - \frac{1}{2^{k-1}}, 1 - \frac{1}{2^k}\right), \ k \in \mathbb{N}, \\ 0 & \text{otherwise.} \end{cases}$$

c) Our last example is a compact operator C on $\ell^\infty(\mathbb{Z}, \mathbf{X})$. Therefore think of a linear functional $f \neq 0$ on this space, such that $f|_{\operatorname{im} P_m} = 0 \ \forall m \in \mathbb{N}$. The existence of such a functional is guaranteed by the Hahn-Banach theorem. If $\mathbf{X} = \mathbb{C}$, one can think, for example, of the extension of the functional that assigns to all convergent sequences their limit at infinity. Now fix an arbitrary $v \in \ell^\infty(\mathbb{Z}, \mathbf{X}) \setminus \{0\}$ and put $C : u \mapsto f(u) \, v$. $\qquad\qquad\square$

A Table of Venn Diagrams of $L(E)$, $L(E, \mathcal{P})$, $K(E, \mathcal{P})$ and $K(E)$

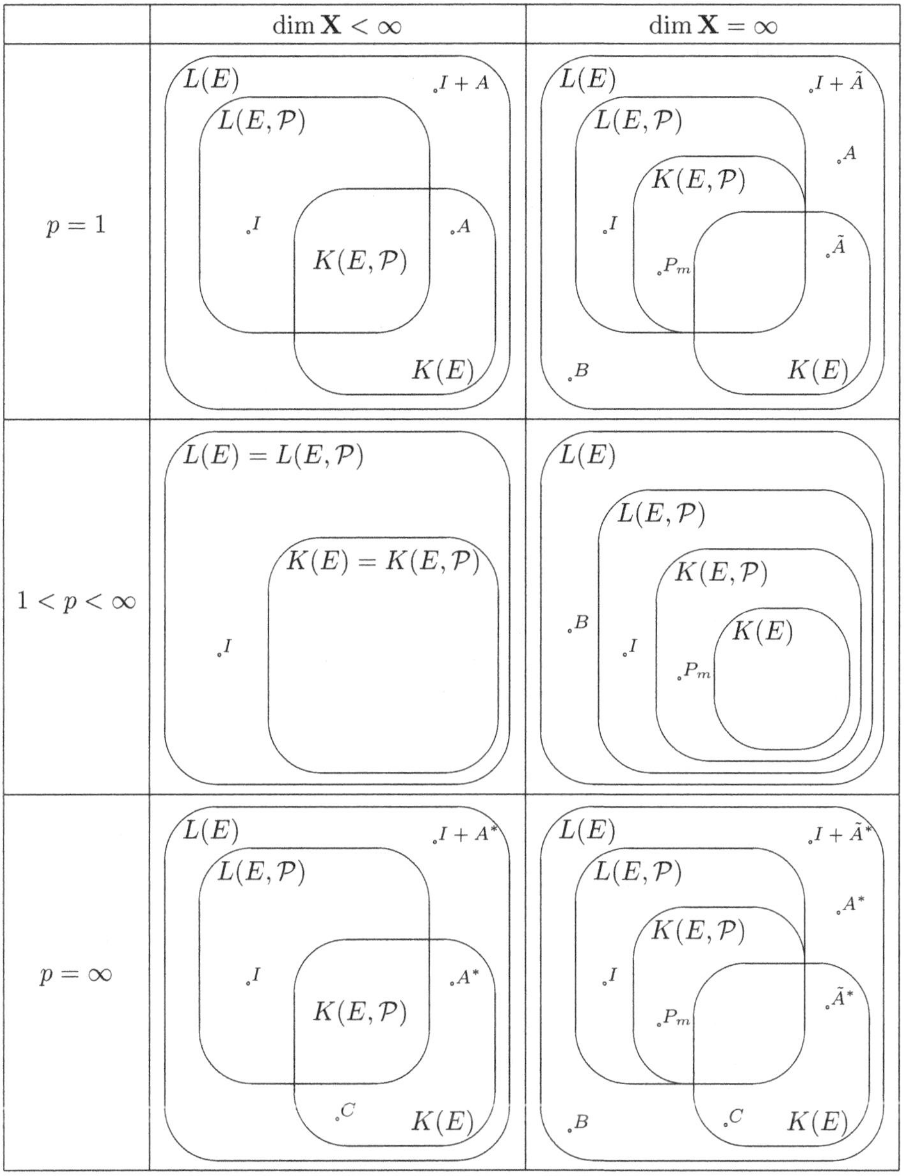

Figure 1: Venn diagrams of $L(E)$, $L(E, \mathcal{P})$, $K(E, \mathcal{P})$ and $K(E)$ depending on $E = \ell^p(\mathbb{Z}^n, \mathbf{X})$.

Note the beauty and simplicity of Figure 1 in the case $1 < p < \infty$, which is called "the perfect case" in [70]. If, in addition, $\dim \mathbf{X} < \infty$, then $K(E, \mathcal{P}) = K(E)$ and consequently, $L(E, \mathcal{P}) = L(E)$, which essentially simplifies many things[4].

---

[4]We will make this statement much more precisely, later in Sections 1.6.3 and 2.3.

1.3.5   Matrix Representation

Given an arbitrary Banach space $\mathbf{X}$ and some $p \in [1, \infty]$, with every bounded linear operator A on $E := \ell^p(\mathbb{Z}^n, \mathbf{X})$, we will associate a matrix

$$[A] = [a_{\alpha\beta}]_{\alpha, \beta \in \mathbb{Z}^n},$$

the entries of which are bounded linear operators on $\mathbf{X}$.

Therefore, for every $\alpha, \beta \in \mathbb{Z}^n$, identify $\operatorname{im} P_{\{\alpha\}}$ with $\mathbf{X}$ in the natural way, and let $a_{\alpha\beta} \in L(\mathbf{X})$ be the operator $P_{\{\alpha\}} A P_{\{\beta\}}|_{\operatorname{im} P_{\{\beta\}}}$, regarded as acting from $\operatorname{im} P_{\{\beta\}} \cong \mathbf{X}$ to $\operatorname{im} P_{\{\alpha\}} \cong \mathbf{X}$. In this sense, we write

$$a_{\alpha\beta} \stackrel{\scriptscriptstyle\frown}{=} P_{\{\alpha\}} A P_{\{\beta\}}, \qquad \alpha, \beta \in \mathbb{Z}^n. \tag{1.16}$$

Consequently, if $u \in E$ has finite support, we get that $v := Au$ is given by

$$v_\alpha = \sum_{\beta \in \mathbb{Z}^n} a_{\alpha\beta} u_\beta, \qquad \alpha \in \mathbb{Z}^n,$$

that is,

$$Au = [A]u, \tag{1.17}$$

by identifying sequences with vectors.

The matrix $[A]$ is referred to as the *matrix representation* of A.

Remark 1.27. a) If $\mathbf{X} = \mathbb{C}$, then the entries $a_{\alpha\beta}$ are just complex numbers. If also $p < \infty$, then $[A]$ is the usual matrix representation of A with respect to the canonical standard basis in ℓ^p.

b) In the continuous case $A \in L(L^p)$, the entry $a_{\alpha\beta}$ of the matrix representation of the discretization (1.3) of A can be identified with $P_{H_\alpha} A P_{H_\beta}|_{\operatorname{im} P_{H_\beta}}$, regarded as acting from $\operatorname{im} P_{H_\beta} \cong L^p(H) = \mathbf{X}$ to $\operatorname{im} P_{H_\alpha} \cong L^p(H) = \mathbf{X}$. In this sense, also in the continuous case, we can think of an operator A on L^p as a matrix of "little" operators $a_{\alpha\beta}$ on $L^p(H)$. We abbreviate this matrix $[A^G]$ by $[A]$ and call it the matrix representation of A. $\square$

Example 1.28. – Integral Operators. If $p \in [1, \infty]$, $n = 1$, and $A \in L(L^p)$ is an *integral operator* on the real line; that is

$$(Au)(x) = \int_{-\infty}^{\infty} k(x, y)\, u(y)\, dy, \qquad x \in \mathbb{R}$$

with an appropriate scalar-valued function k on $\mathbb{R}^2$, referred to as the *kernel function* of A, then, for all $i, j \in \mathbb{Z}$,

$$(P_{H_i} A P_{H_j} u)(x) = \int_j^{j+1} k(x, y)\, u(y)\, dy, \qquad x \in [i, i+1]$$

is an integral operator acting from $L^p([j, j+1])$ to $L^p([i, i+1])$. By shifting x and y to $[0,1]$, we get that the entries of $[A] = [a_{ij}]_{i,j \in \mathbb{Z}}$ are the "little" integral operators

$$(a_{ij} u)(x) = \int_0^1 k(x+i, y+j) u(y) \, dy, \qquad x \in [0,1]$$

on the interval $[0,1]$. $\qquad\qquad\qquad\qquad\qquad\qquad\qquad\qquad\qquad\qquad\qquad$ □

Proposition 1.29.   a) *For arbitrary operators $A, B \in L(E)$ with $[A] = [a_{\alpha\beta}]_{\alpha,\beta\in\mathbb{Z}^n}$ and $[B] = [b_{\alpha\beta}]_{\alpha,\beta\in\mathbb{Z}^n}$, one has*

$$[A+B] = [A] + [B] \qquad and \qquad [AB] = [A][B],$$

where $[A] + [B] := [a_{\alpha\beta} + b_{\alpha\beta}]_{\alpha,\beta\in\mathbb{Z}^n}$ and $[A][B] := [\sum_{\gamma\in\mathbb{Z}^n} a_{\alpha\gamma} b_{\gamma\beta}]_{\alpha,\beta\in\mathbb{Z}^n}$.

b) *If $A \in L(E)$ with $p < \infty$ and $[A] = [a_{\alpha\beta}]$, then $[A^*] = [a^*_{\beta\alpha}]$.*

c) *Take a sequence (A_m) of operators on E and let $[A_m] = [a^{(m)}_{\alpha\beta}]$ denote their matrix representations.*

If $A_m \rightrightarrows 0$, then $a^{(m)}_{\alpha\beta} \rightrightarrows 0$ uniformly with respect to α and β as $m \to \infty$.

If $A_m \to 0$, then $a^{(m)}_{\alpha\beta} \to 0$ for all α and β as $m \to \infty$.

Proof. a) From (1.17) we get that, for every finitely supported $u \in E$,

$$[A+B]u = (A+B)u = Au + Bu = [A]u + [B]u = ([A] + [B])u$$
$$\text{and} \qquad [AB]u = ABu = A(Bu) = [A](Bu) = [A][B]u$$

hold, which proves the claim.

b) From $P_{\{\alpha\}} A^* P_{\{\beta\}} = P^*_{\{\alpha\}} A^* P^*_{\{\beta\}} = (P_{\{\beta\}} A P_{\{\alpha\}})^*$ and (1.16) we get that the entry in position (α, β) of $[A^*]$ equals $a^*_{\beta\alpha}$.

c) If $A_m \rightrightarrows 0$, then, by (1.16), $\|a^{(m)}_{\alpha\beta}\| = \|P_{\{\alpha\}} A_m P_{\{\beta\}}\| \le \|A_m\| \to 0$ as $m \to \infty$. Now suppose $A_m \to 0$ as $m \to \infty$. Choose arbitrary $\alpha, \beta \in \mathbb{Z}^n$ and an $x \in \mathbf{X}$. Define $u \in \ell^p(\mathbb{Z}^n, \mathbf{X})$ by $u_\beta = x$ and $u_\gamma = 0$ for $\gamma \neq \beta$. Since $A_m \to 0$, we get $v^{(m)} := A_m u \to 0$ as $m \to \infty$. From $\|v^{(m)}_\alpha\|_{\mathbf{X}} \le \|v^{(m)}\|$ we get that also the α-th component $v^{(m)}_\alpha = a_{\alpha\beta} x$ tends to zero in $\mathbf{X}$ as $m \to \infty$. Since α, β and x were chosen arbitrarily, we have proven the claim. $\qquad\qquad$ □

From (1.16) it is clear that, for every $A \in L(E)$, there is exactly one matrix representation $[A]$. A much more delicate question is whether also $[A]$ uniquely determines the operator A or not. From Proposition 1.29 a) we conclude that $A \mapsto [A]$ is a linear mapping from $L(E)$ to the space of all $L(\mathbf{X})$-valued matrices over $\mathbb{Z}^n \times \mathbb{Z}^n$. Now the question is about the kernel of that mapping. But first we insert an auxiliary result.

Lemma 1.30.

a) $P_{\{\alpha\}}A = 0 \ \forall \alpha \in \mathbb{Z}^n \quad \Longleftrightarrow \quad P_m A = 0 \ \forall m \in \mathbb{N} \quad \Longleftrightarrow \quad A = 0.$

b) $A P_{\{\beta\}} = 0 \ \forall \beta \in \mathbb{Z}^n \quad \Longleftrightarrow \quad A P_m = 0 \ \forall m \in \mathbb{N} \quad \Longleftarrow \quad A = 0.$

Proof. a)If $P_{\{\alpha\}}A = 0 \ \forall \alpha \in \mathbb{Z}^n$, then also $P_m A = \sum_{|\alpha| \le m} P_{\{\alpha\}}A = 0 \ \forall m \in \mathbb{N}$. Now suppose $A \neq 0$. Then there is an $u \in E$ such that $Au \neq 0$, i.e. $(Au)_\alpha \neq 0$ for some $\alpha \in \mathbb{Z}^n$. But then $P_m Au \neq 0$ and hence, $P_m A \neq 0$ for all $m \ge |\alpha|$. Contradiction. Finally, $A = 0$ obviously implies the other two conditions.

b) If $A P_{\{\beta\}} = 0 \ \forall \beta \in \mathbb{Z}^n$, then also $A P_m = \sum_{|\beta| \le m} A P_{\{\beta\}} = 0 \ \forall m \in \mathbb{N}$. Conversely, if $A P_m = 0$, then also $A P_{\{\beta\}} = A P_m P_{\{\beta\}} = 0$ if $|\beta| \le m$. Finally, $A = 0$ obviously implies the other conditions.   □

The first two conditions in Lemma 1.30 b) turn out to be equivalent to $[A] = 0$.

Proposition 1.31.   a) *The kernel of the mapping*

$$[\,.\,] \ : \ A \ \mapsto \ [A]$$

consists of the operators A for which $A P_{\{\beta\}} = 0$ for all $\beta \in \mathbb{Z}^n$, or, which is equivalent, $A P_m = 0$ for all $m \in \mathbb{N}$.

b) *If $p < \infty$, then the mapping $[\,.\,] : A \mapsto [A]$ is one-to-one, i.e. $\ker[\,.\,] = \{0\}$.*

c) *For $p = \infty$, $\ker[\,.\,]$ is an infinite-dimensional subspace of $L(E)$ which intersects with $L(E, \mathcal{P})$ and with $\mathcal{S}$ in 0 only.*

Proof. a) $[A] = 0$ is equivalent to $P_{\{\alpha\}} A P_{\{\beta\}} = 0$ for all $\alpha, \beta \in \mathbb{Z}^n$. From Lemma 1.30 a) we get that this is equivalent to $A P_{\{\beta\}} = 0$ for all $\beta \in \mathbb{Z}^n$, which is moreover equivalent to $A P_m = 0$ for all $m \in \mathbb{N}$ by Lemma 1.30 b).

b) If $p < \infty$, then $P_m \to I$ strongly as $m \to \infty$. If $A \in \ker[\,.\,]$, then we know from a) that $A P_m = 0$ for all $m \in \mathbb{N}$. Passing to the strong limit as $m \to \infty$, we get $A = 0$.

c) For $p = \infty$, there are indeed operators $A \neq 0$ with $[A] = 0$. A standard example is the operator $u \mapsto f(u)v$ in Example 1.26 c). If this functional f is fixed and v runs through an infinite set of linearly independent elements in E, then we get an infinite variety of linearly independent operators in $\ker[\,.\,]$.

Now suppose that $A \in L(E, \mathcal{P})$ is in $\ker[\,.\,]$. From $A P_m = 0$ we get that $A Q_m = A$ for all $m \in \mathbb{N}$. The first condition in (1.12) then shows that, for every $k \in \mathbb{N}$,

$$P_k A \ = \ P_k A Q_m \rightrightarrows 0 \qquad \text{as} \qquad m \to \infty,$$

i.e. $P_k A = 0$ for every $k \in \mathbb{N}$. From Lemma 1.30 a) we get $A = 0$.

Finally suppose that $A \in \mathcal{S}$ is in $\ker[\,.\,]$, and put $B = A^{\triangleleft}$. From Proposition 1.29 b) we get that all entries of $[B]$ are zero as well. But since B is an operator on $\ell^1(\mathbb{Z}^n, \mathbf{Y})$ with $\mathbf{Y}^* = \mathbf{X}$, we know from b) that $B = 0$, and consequently $A = 0$.   □

So in the case $p = \infty$, we have two completely different settings which both guarantee that for a given matrix $[A]$ there is only one operator A within that setting:

$$\text{(i)} \quad A \in L(E, \mathcal{P})$$
$$\text{(ii)} \quad A \in \mathcal{S}$$

Both conditions (i) and (ii) are sufficient but not necessary for the uniqueness of $[A] \mapsto A$. Moreover, neither of the two conditions implies the other one. To see this,

- firstly, think of a generalized multiplication operator $\hat{M}_b$ where some component of the sequence b has no pre-adjoint operator. Then $A = \hat{M}_b$ is subject to condition (i) but not to (ii).

- Secondly, let $(Au)_\alpha := u_0$ for all $\alpha \in \mathbb{Z}^n$. Then A is subject to condition (ii) since A is the adjoint operator of Example 1.26 a), but A is not subject to (i) since the second condition in (1.12) is violated.

For a given space $E = \ell^p(\mathbb{Z}^n, \mathbf{X})$, we put

$$M(E) := \begin{cases} L(E) & \text{if } p < \infty, \\ L(E, \mathcal{P}) \cup \mathcal{S} & \text{if } p = \infty, \end{cases}$$

which makes $M(E)$ some fairly large set of operators for which we know that no two elements have the same matrix representation. By restricting ourselves to operators in $M(E)$, we will gently walk around certain difficulties that can arise if $p = \infty$. The uniqueness of the operator behind a given matrix will turn out to be only one example of such a problem.

Although the entries of our matrices $[A]$ are indexed slightly more complicated than in the matrices that one usually has in mind, we will speak about rows, columns and diagonals of $[A]$ in the usual way:

Definition 1.32. *For an operator $A \in L(E)$ with matrix representation $[A] = [a_{\alpha\beta}]$, we refer to the sequence $(a_{\alpha\beta})_{\beta\in\mathbb{Z}^n}$ as the α-th row, to $(a_{\alpha\beta})_{\alpha\in\mathbb{Z}^n}$ as the β-th column, and to $(a_{\beta+\gamma,\beta})_{\beta\in\mathbb{Z}^n}$ as the γ-th diagonal of $[A]$ – or simply, of A. As usual, the 0-th diagonal is also called the main diagonal. Moreover, we will refer to $(a_{\alpha,-\alpha})_{\alpha\in\mathbb{Z}^n}$ as the cross diagonal of A.*

Example 1.33. a) To discuss some simple examples, first let $A = \hat{M}_b$ be a generalized multiplication operator. Then $[A]$ is a diagonal matrix, i.e. it is supported on the main diagonal only, with $a_{\alpha\alpha} = b_\alpha$ where $b = (b_\alpha)$ is the symbol of A. For operators in $M(E)$, also the reverse statement is true: A is a generalized multiplication operator if and only if $[A]$ is a diagonal matrix.

b) As a second example, put $A = V_\gamma$ with $\gamma \in \mathbb{Z}^n$. Then the matrix $[A]$ is supported on the γ-th diagonal only. Precisely, $a_{\alpha\beta}$ is $I_{\mathbf{X}}$, the identity operator on $\mathbf{X}$, if $\alpha - \beta = \gamma$, and it is 0 otherwise.

c) Now we combine the two previous examples and study $A = \hat{M}_b V_\gamma$. From Proposition 1.29 a) and Examples 1.33 a) and b), we get that $[A] = [\hat{M}_b][V_\gamma]$ is supported on the γ-th diagonal only, where $a_{\alpha\beta} = b_\alpha$ if $\alpha - \beta = \gamma$, and $a_{\alpha\beta} = 0$ otherwise. In analogy to a), also here the reverse statement is true for operators in $M(E)$: A is of the form $\hat{M}_b V_\gamma$ if and only if $[A]$ is only supported on the γ-th diagonal.

d) As a final example, put $A = P_U$ with $U \subset \mathbb{Z}^n$. Then A is a very simple example of a generalized multiplication operator. So $[A]$ is a diagonal matrix again, where $a_{\alpha\beta} = I_\mathbf{X}$ if $\alpha = \beta \in U$, and $a_{\alpha\beta} = 0$ otherwise.   □

1.3.6   Band- and Band-dominated Operators

One usually regards a matrix $[a_{ij}]$ with i, j running through $\{1, \ldots, m\}$ or $\mathbb{N}$ or even through $\mathbb{Z}$ as a band matrix if its entries a_{ij} vanish whenever $|i - j|$ is larger than some number, which is then called the band-width of the matrix. As introduced in Section 1.3.5, for the investigation of operators on $E = \ell^p(\mathbb{Z}^n, \mathbf{X})$, we are interested in the study of matrices $[a_{\alpha\beta}]$ which are indexed over $\mathbb{Z}^n \times \mathbb{Z}^n$.

We call $[a_{\alpha\beta}]_{\alpha,\beta \in \mathbb{Z}^n}$ a *band matrix* of *band-width* w if all entries $a_{\alpha\beta}$ with $|\alpha - \beta| > w$ vanish, or, what is equivalent, if the matrix is only supported on the γ-th diagonals with $|\gamma| \leq w$. Now we are looking for operators $A \in L(E)$ with such a matrix representation. From Example 1.33 c) we know that for operators of the form

$$A = \sum_{|\gamma| \leq w} \hat{M}_{b^{(\gamma)}} V_\gamma \tag{1.18}$$

with $b^{(\gamma)} \in \ell^\infty(\mathbb{Z}^n, L(\mathbf{X}))$ for every $\gamma \in \mathbb{Z}^n$ involved in the summation, the matrix representation $[A]$ is a band matrix with band-width w. Example 1.33 c) also tells us that the operators $A \in M(E)$ with a band matrix representation $[A]$ of band-width w are exactly those of the form (1.18). We will therefore use equation (1.18) to define one of the most important classes of operators in this book:

Definition 1.34. *An operator A is called a band operator of band-width w if it is of the form* (1.18). *The set of all band operators is denoted by* BO.

Remark 1.35. A band operator acts boundedly on all spaces $E = \ell^p(\mathbb{Z}^n, \mathbf{X})$ with $p \in [1, \infty]$. Also note that it is automatically contained in $M(E)$. There are operators A outside of $M(E)$ for which $[A]$ is a band matrix as well but which we do not regard as band operators, for instance the operator in Example 1.26 c).   □

Proposition 1.36. *For a bounded linear operator A on $E = \ell^p(\mathbb{Z}^n, \mathbf{X})$ with matrix representation $[A] = [a_{\alpha\beta}]$, the following three properties are equivalent:*

(i) *A is a band operator with band-width w,*

(ii) *$A \in M(E)$ and $a_{\alpha\beta} = 0$ for $|\alpha - \beta| > w$,*

(iii) *$P_V A P_U = 0$ for all sets $U, V \subset \mathbb{Z}^n$ with $\mathrm{dist}\,(U, V) > w$.*

Proof. (i)⇒(iii). Suppose that A is of the form (1.18) and that $U, V \subset \mathbb{Z}^n$ have a distance larger than w. Then

$$P_V A P_U = \sum_{|\gamma| \le w} P_V \hat{M}_{b(\gamma)} V_\gamma P_U = \sum_{|\gamma| \le w} \hat{M}_{b(\gamma)} P_V P_{\gamma+U} V_\gamma = 0$$

since $P_V P_{\gamma+U} = P_{V \cap (\gamma+U)} = P_\varnothing = 0$ if $|\gamma| \le w < \mathrm{dist}\,(U, V)$.

(iii)⇒(ii). Take arbitrary $\alpha, \beta \in \mathbb{Z}^n$ with $|\alpha - \beta| > w$. Put $U := \{\beta\}$ and $V := \{\alpha\}$, and remember (1.16) to see that

$$\|a_{\alpha\beta}\| = \|P_{\{\alpha\}} A P_{\{\beta\}}\| = \|P_V A P_U\| = 0.$$

Moreover, (iii) implies $P_k A Q_m = 0 = Q_m A P_k$ for all $m > k + w$, which shows that $A \in L(E, \mathcal{P}) \subset M(E)$.

(ii)⇒(i) follows immediately from the discussion in Example 1.33 c). $\qquad\square$

It is easily seen that BO is a subalgebra of $L(E)$, as sums and products of band matrices are band matrices again. Unfortunately, this subalgebra is not closed! So it is a very natural desire to pass to the closure of BO which is a Banach algebra.

Definition 1.37. *For every $p \in [1, \infty]$, let BDO^p refer to the closure, with respect to the operator norm in $L(\ell^p(\mathbb{Z}^n, \mathbf{X}))$, of the algebra BO of band operators. The elements of BDO^p are called band-dominated operators.*

We will moreover refer to the collection $\{b^{(\gamma)}\}_{|\gamma| \le w}$ as *the diagonals* or *the coefficients* of the band operator A in (1.18). If F is a subset of $\ell^\infty(\mathbb{Z}^n, L(\mathbf{X}))$ and $A \in \mathrm{BDO}^p$ is the norm limit of a sequence of band operators with coefficients in F, then we will say that A is a *band-dominated operator with coefficients in F*.

Remark 1.38. BO is the smallest algebra that contains all generalized multiplication operators and shift operators. Its elements are finite sum-products of these objects. BDO^p is the smallest Banach algebra that contains all generalized multiplication operators and shifts. In short:

$$\mathrm{BO} = \mathrm{alg}_{L(\ell^p(\mathbb{Z}^n, \mathbf{X}))} \left\{ \hat{M}_b, V_\alpha : b \in \ell^\infty(\mathbb{Z}^n, L(\mathbf{X})), \alpha \in \mathbb{Z}^n \right\},$$

$$\mathrm{BDO}^p = \mathrm{closalg}_{L(\ell^p(\mathbb{Z}^n, \mathbf{X}))} \left\{ \hat{M}_b, V_\alpha : b \in \ell^\infty(\mathbb{Z}^n, L(\mathbf{X})), \alpha \in \mathbb{Z}^n \right\}.$$

Sometimes BO and BDO^p are also referred to as the algebra and the Banach algebra, respectively, that are *generated by* generalized multiplications and shifts. In that sense, we wrote about $\hat{M}_b$ and V_α as the basic building stones or atoms of the class BDO^p in Section 1.3.1, and we will do so in what follows.

Note that, as the notation already suggests, the class BO does not depend on the Lebesgue exponent $p \in [1, \infty]$ of the underlying space, while BDO^p heavily does! $\qquad\square$

Example 1.39. – Laurent Operators. We have a short intermezzo on Laurent operators. These are operators, the matrix representation of which is constant along every diagonal. Let $p = 2$, $n = 1$, $\mathbf{X} = \mathbb{C}$, and fix a function $a \in L^\infty(\mathbb{T})$. The operator

$$A = \sum_{k=-\infty}^{\infty} a_k V_k \qquad \text{i.e.} \qquad [A] = \begin{pmatrix} \ddots & \ddots & \ddots & & \ddots \\ \ddots & a_0 & a_{-1} & a_{-2} & \ddots \\ \ddots & a_1 & a_0 & a_{-1} & \ddots \\ \ddots & a_2 & a_1 & a_0 & \ddots \\ & \ddots & \ddots & \ddots & \ddots \end{pmatrix}$$

on $E = \ell^2$ is called *Laurent operator*, where $a_k \in \mathbb{C}$ are the Fourier coefficients

$$a_k = \frac{1}{2\pi} \int_0^{2\pi} a(e^{i\theta}) e^{-ik\theta}\, d\theta, \qquad k \in \mathbb{Z}$$

of the function a on the unit circle $\mathbb{T}$, which is referred to as the *symbol* of $A =:$ $L(a)$.

It is readily shown that, via the Fourier isomorphism between ℓ^2 and $L^2(\mathbb{T})$, the Laurent operator $L(a)$ on ℓ^2 corresponds to the operator of multiplication by a in $L^2(\mathbb{T})$. As a consequence, one gets that

$$
\begin{aligned}
L(a) \in L(\ell^2) &\iff a \in L^\infty(\mathbb{T}), \\
\|L(a)\| &= \|a\|_\infty, & a \in L^\infty(\mathbb{T}), & \qquad (1.19) \\
L(a) + L(b) &= L(a + b), & a, b \in L^\infty(\mathbb{T}), & \qquad (1.20) \\
L(a)L(b) &= L(ab), & a, b \in L^\infty(\mathbb{T}), & \\
L(a) \text{ is invertible in } L(\ell^2) &\iff a \text{ is invertible in } L^\infty(\mathbb{T}). &
\end{aligned}
$$

As a consequence of (1.19) and (1.20), we moreover get that

$$
\begin{aligned}
L(a) \in \mathrm{BO} &\iff \#\{k : a_k \neq 0\} < \infty \iff a \text{ is trigonometric polynomial}, \\
L(a) \in \mathrm{BDO}^2 &\iff a \in \mathrm{clos}_{L^\infty(\mathbb{T})}\{\text{trigonometric polynomials}\} = C(\mathbb{T}).
\end{aligned}
$$

Note that the class M^p of all symbols $a \in L^\infty(\mathbb{T})$, for which $L(a) \in L(\ell^p)$ holds, decreases as soon as $p \in [1, \infty]$ differs from 2. The same is true for the set of all $a \in M^p$ for which $L(a) \in \mathrm{BDO}^p$. For every $p \in [1, \infty]$, the set M^p is a Banach subalgebra of $M^2 = L^\infty(\mathbb{T})$, equipped with the norm

$$\|a\|_{M^p} := \|L(a)\|_{L(\ell^p)}. \qquad (1.21)$$

Moreover, it holds that

$$M^1 \subset M^{p_1} \subset M^{p_2} \subset M^2 = L^\infty(\mathbb{T})$$

if $1 \leq p_1 \leq p_2 \leq 2$, and $M^p = M^q$ if $1/p + 1/q = 1$, including the equality $M^1 = M^\infty$. From (1.21) it follows that

$$L(a) \in \mathrm{BDO}^p \quad \Longleftrightarrow \quad a \in \mathrm{clos}_{M^p}\{\text{trigonometric polynomials}\} \subset C(\mathbb{T}).$$

It follows that $L(a) \in \mathrm{BDO}^{p_1}$ implies $L(a) \in \mathrm{BDO}^{p_2}$ for all $1 \leq p_1 \leq p_2 \leq 2$, and that $L(a) \in \mathrm{BDO}^p$ if and only if $L(a) \in \mathrm{BDO}^q$ with $1/p + 1/q = 1$. In particular, the latter holds with $p = 1$ and $q = \infty$ if and only if $a \in \mathcal{W}(\mathbb{T})$; that is, $\sum |a_k| < \infty$, which is a proper subclass of $C(\mathbb{T})$.

So this is certainly an example of the dependence of BDO^p on p. $\qquad\square$

Remark 1.40. The impression that, for every $A \in \mathrm{BDO}^p$, it holds that $A_m \rightrightarrows A$ where $[A_m]$ is just the restriction of $[A]$ to a finite number of diagonals, is false in general!

As a counter-example, take $p = 2$, $n = 1$, $\mathbf{X} = \mathbb{C}$, and $A = L(a)$, where $a \in C(\mathbb{T})$ is such that the sequence of partial Fourier sums

$$t \in \mathbb{T} \quad \mapsto \quad \sum_{k=-m}^{m} a_k \, t^k$$

is not uniformly convergent to a as $m \to \infty$. Consequently, by (1.19),

$$A_m = \sum_{k=-m}^{m} a_k V_k \quad \not\rightrightarrows \quad A \qquad \text{as} \qquad m \to \infty.$$

By Fejer's theorem [39, Theorem 3.1] we know that, however, the Fejer Cesaro means uniformly converge to a; that is

$$t \mapsto \sum_{k=-m}^{m} \left(1 - \frac{|k|}{m+1} \right) a_k \, t^k \qquad \to \qquad a$$

in the norm of $L^\infty(\mathbb{T})$ as $m \to \infty$, showing that $\tilde{A}_m \rightrightarrows A$ with

$$\tilde{A}_m = \sum_{k=-m}^{m} \left(1 - \frac{|k|}{m+1} \right) a_k V_k \in \mathrm{BO}. \qquad\square$$

We will now prepare Theorem 1.42 which gives some equivalent characterizations of BDO^p. One of these characterizations corresponds to condition (iii) in Proposition 1.36. The other characterizations have something to do with the class of bounded and uniformly continuous functions:

Definition 1.41. *By* BC *we denote the set of all bounded and continuous complex-valued functions on $\mathbb{R}^n$. Moreover, let* BUC *denote the subset of all uniformly continuous functions among* BC.

Given a function $\varphi \in L^\infty$ and a vector $t = (t_1, \ldots, t_n) \in \mathbb{R}^n$, we define the function φ_t at the point $x = (x_1, \ldots, x_n) \in \mathbb{R}^n$ as

$$\varphi_t(x) := \varphi(t_1 x_1, \ldots, t_n x_n). \tag{1.22}$$

Moreover, if $r \in \mathbb{R}^n$, then we put $\varphi_{t,r}(x) := \varphi_t(x - r)$ for all $x \in \mathbb{R}^n$.

Now take an arbitrary function $\varphi \in \mathrm{BUC}$ and watch the functions φ_t as $t \to 0$; that is, blowing up the argument of φ such that the graph becomes flatter and flatter. If we compare that function φ_t with a shifted copy $V_\kappa \varphi_t$ for fixed $\kappa \in \mathbb{Z}^n$, then it clearly follows from $\varphi \in \mathrm{BUC}$, that the difference $\varphi_t - V_\kappa \varphi_t$ uniformly tends to zero the more we blow up the argument, i.e. as $t \to 0$.

This observation shows that, for every $\varphi \in \mathrm{BUC}$ and every fixed $\kappa \in \mathbb{R}^n$,

$$\|M_{\varphi_t} V_\kappa - V_\kappa M_{\varphi_t}\| = \|M_{\varphi_t} - V_\kappa M_{\varphi_t} V_{-\kappa}\| = \|M_{\varphi_t - V_\kappa \varphi_t}\| = \|\varphi_t - V_\kappa \varphi_t\|_\infty \tag{1.23}$$

tends to zero as $t \to 0$. We will now change our point of view and regard this convergence process as a property of the shift operator: $A = V_\kappa$ is subject to

$$[M_{\varphi_t}, A] \rightrightarrows 0 \qquad \text{as} \qquad t \to 0 \tag{1.24}$$

for all $\varphi \in \mathrm{BUC}$. Roughly speaking, property (1.24) says that A commutes increasingly better with M_{φ_t} the more we blow up the function φ_t.

Obviously, property (1.24) is written down for the continuous case, $A \in L(L^p)$, because this is the natural playground for functions $\varphi_t \in \mathrm{BUC}$ and the operator M_{φ_t} of multiplication by such. But we will transport this property to the discrete case $A \in L(\ell^p(\mathbb{Z}^n, \mathbf{X}))$ by finding the discrete analogue of M_{φ_t}.

For $f \in \mathrm{BC}$, let $\hat{M}_f$ denote the generalized multiplication operator $\hat{M}_f := \hat{M}_b$ with $b = (b_\alpha)$, where $b_\alpha = f(\alpha) I_\mathbf{X}$, i.e. $\hat{M}_f$ acts on $u \in \ell^p(\mathbb{Z}^n, \mathbf{X}))$ by

$$(\hat{M}_f u)_\alpha = f(\alpha) u_\alpha \qquad \forall \alpha \in \mathbb{Z}^n. \tag{1.25}$$

Note that the operator $\hat{M}_f$ evaluates the function $f \in \mathrm{BC}$ at the integer points $\alpha \in \mathbb{Z}^n$ only.

Then we study the set of all operators A on $\ell^p(\mathbb{Z}^n, \mathbf{X})$ such that

$$[\hat{M}_{\varphi_t}, A] \rightrightarrows 0 \qquad \text{as} \qquad t \to 0 \tag{1.26}$$

holds for all $\varphi \in \mathrm{BUC}$. Elementary computations show that this set is a Banach algebra. Above we have convinced ourselves that it contains all shift operators. Moreover, trivially, all generalized multiplication operators are subject to (1.26). Consequently, every operator $A \in \mathrm{BDO}^p$ is so. Much more surprisingly, it turns out that the set of bounded and linear operators which are subject to (1.26) **equals** BDO^p. Here we go:

Theorem 1.42. *For every $p \in [1, \infty]$, every Banach space $\mathbf{X}$ and every bounded linear operator A on $E = \ell^p(\mathbb{Z}^n, \mathbf{X})$, the following properties are equivalent:*

(i) *A is a band-dominated operator.*

(ii) *$P_V A P_U \rightrightarrows 0$ as $\operatorname{dist}(U, V) \to \infty$ in the following sense:*

$$\forall \varepsilon > 0 \ \exists d > 0: \quad \|P_V A P_U\| < \varepsilon \text{ for all } U, V \subset \mathbb{Z}^n \ : \ \operatorname{dist}(U, V) > d.$$

(iii) *$[\hat{M}_{\varphi_{t,r}}, A] \rightrightarrows 0$ as $t \to 0$, uniformly w.r.t. $r \in \mathbb{R}^n$, for all $\varphi \in \mathrm{BUC}$.*

(iv) *(1.26), i.e. $[\hat{M}_{\varphi_t}, A] \rightrightarrows 0$ as $t \to 0$ holds for all $\varphi \in \mathrm{BUC}$.*

(v) *$A \in M'(E)$ and (1.26) holds for $\varphi(x_1, \ldots, x_n) = \exp i(x_1 + \cdots + x_n)$.*

Here $M'(E)$ denotes the following subset of $M(E)$,

$$M'(E) := \begin{cases} L(E) & \text{if } p < \infty, \\ L(E, \mathcal{P}) \cup \mathcal{S} & \text{if } p = \infty \text{ and } \mathbf{X} = L^\infty(H), \\ L(E, \mathcal{P}) & \text{otherwise.} \end{cases}$$

This theorem is a conglomeration of Theorem 2.1.6 in [70] and Proposition 1.16 in [50] (alias Proposition 2.13 in [47]) with the slight but convenient modification of moving the condition $A \in M'(E)$ from the top of the theorem to condition (v) since it turns out to be redundant in (i)–(iv). In [41] KURBATOV gives an example of an operator A on $\ell^\infty(\mathbb{Z})$ which shows that the condition $A \in M'(E)$ is not redundant in (v).

Proof. (i)$\Rightarrow$(ii)$\Rightarrow$(iii)$\Rightarrow$(iv) is shown in the proof of Theorem 2.1.6 in [70].

(iv)$\Rightarrow$(v). Put $E = \ell^p(\mathbb{Z}^n, \mathbf{X})$. In Proposition 2.4 of [68] it is shown that AT and TA are in $K(E, \mathcal{P})$ for all $T \in K(E, \mathcal{P})$ and A being subject to (iv). This clearly shows that (iv) implies $A \in L(E, \mathcal{P}) \subset M'(E)$ by Definition 1.18. Obviously, (iv) also implies the validity of condition (1.26) for the particular function $\varphi \in \mathrm{BUC}$ mentioned in (v).

(v)$\Rightarrow$(i). This implication is proven in Theorem 2.1.6 of [70] for $p < \infty$ and for all $A \in L(\ell^\infty(\mathbb{Z}^n, \mathbf{X}), \mathcal{P})$. For the missing case, $p = \infty$, $\mathbf{X} = L^\infty(H)$, $A \in \mathcal{S}$, it is proven in [50], Proposition 1.16 and also in [47], Proposition 2.13. The key in the latter case is a combination of arguments from [41] and [80]. $\qquad\square$

Finally, we define a fairly large and especially well-behaved subalgebra of BDO^p – the so-called Wiener algebra. Again abbreviate $\ell^p(\mathbb{Z}^n, \mathbf{X})$ by E, where n and $\mathbf{X}$ are fixed, and only p shall vary in $[1, \infty]$ if we say so. For a band operator A of the form (1.18), one clearly has

$$\|A\|_{L(E)} = \left\| \sum_{|\gamma| \le w} \hat{M}_{b^{(\gamma)}} V_\gamma \right\|_{L(E)} \le \sum_{\gamma \in \mathbb{Z}^N} \|b^{(\gamma)}\|_\infty =: \|A\|_w \qquad (1.27)$$

for all $p \in [1, \infty]$, where we put $b^{(\gamma)} = 0$ if $|\gamma|$ exceeds the band-width w of A.

Definition 1.43. *By* $\mathcal{W}$ *we denote the closure of* BO *with respect to* $\|.\|_{\mathcal{W}}$. *Equipped with* $\|.\|_{\mathcal{W}}$, *this set turns out to be a Banach algebra, henceforth called the Wiener algebra.*

From Definition 1.43 and inequality (1.27) we get that $\mathcal{W}$ is contained in BDO^p for all $p \in [1, \infty]$,

$$\text{BO} \subset \mathcal{W} \subset \bigcap_{1 \leq p \leq \infty} \text{BDO}^p. \tag{1.28}$$

In Example 1.49 d) and e) and in Figure 2 on page 28 we will see that both inclusions in (1.28) are proper. Like BO, the class $\mathcal{W}$ does not depend on $p \in [1, \infty]$. An operator $A \in \mathcal{W}$ can be thought of as acting on any of the spaces $E = \ell^p(\mathbb{Z}^n, \mathbf{X})$, $p \in [1, \infty]$, where the inequality

$$\|A\|_{L(E)} \leq \|A\|_{\mathcal{W}} \tag{1.29}$$

from (1.27) extends to all $A \in \mathcal{W}$.

Remark 1.44. Suppose $A \in \mathcal{W}$, and $[A] = [a_{\alpha\beta}]_{\alpha,\beta\in\mathbb{Z}^n}$. Unlike for arbitrary band-dominated operators A (see Remark 1.40), for operators $A \in \mathcal{W}$, the sequence of band operators A_m with $[A_m] = [a_{\alpha\beta}]_{|\alpha-\beta|\leq m}$ converges to A in the $\mathcal{W}$-norm, and hence in the norm of $L(E)$ for all $p \in [1, \infty]$.   □

Example 1.45. – Convolution Operators on $\mathbb{R}$**.** We will have another look at the continuous case $A \in L(L^p)$ with $n = 1$ and $p \in [1, \infty]$, where A is a so-called *convolution operator* on the real line; that is an integral operator as in Example 1.28, the kernel function $k(x, y)$ of which only depends on the difference of x and y. So let

$$(Au)(x) = \int_{-\infty}^{\infty} \kappa(x - y)\, u(y)\, dy, \qquad x \in \mathbb{R},$$

where $\kappa \in L^1$. From Example 1.28 we know that the entries of $[A] = [a_{ij}]$ are

$$(a_{ij}\, u)(x) = \int_0^1 \kappa(x - y + i - j)\, u(y)\, dy, \qquad x \in [0, 1],$$

for all $i, j \in \mathbb{Z}$, showing that also a_{ij} depends on the difference $i - j$ only. An elementary calculation for $p = 1$ and $p = \infty$, together with Riesz Thorin interpolation (see [39, IV.1.4] or [84, V.1]) shows that

$$\|a_{ij}\| \leq \sup_{c\in[d-1,\,d]} \|\kappa|_{[c,\,c+1]}\|_{L^1}$$

for all $d = i - j \in \mathbb{Z}$.

Together with $\kappa \in L^1$, this clearly yields $A^G \in \mathcal{W}$ for the discretization (1.3) of A, showing that A is band-dominated for every choice of $p \in [1, \infty]$, and that

$$\|A\|_{L(L^p)} = \|A^G\|_{L(\ell^p(\mathbb{Z},\mathbf{X}))} \leq \|A^G\|_{\mathcal{W}} \leq 2\|\kappa\|_{L^1}$$

holds with $\mathbf{X} = L^p([0,1])$. Young's inequality [72] shows that the factor 2 even can be omitted in the inequality $\|A\|_{L(L^p)} \le 2\|\kappa\|_{L^1}$. $\qquad\square$

The following is a very deep and vital result in the theory of the operator classes introduced and studied here. As before, we abbreviate $\ell^p(\mathbb{Z}^n, \mathbf{X})$ by E.

Proposition 1.46.   a) $L(E, \mathcal{P})$ *is inverse closed in* $L(E)$.

b) BDO^p *is inverse closed in* $L(E)$.

c) $\mathcal{W}$ *is inverse closed in* $L(E)$.

Proof. a) and c) are both very deep observations with very technical proofs, for which we frankly refer to [70], Theorem 1.1.9 and Theorem 2.5.2, respectively. (Note that our $\mathcal{P}$ is a uniform approximate projection in the sense of [70].)
b) This follows immediately from Theorem 1.42 (iv): If $A \in BDO^p$ is invertible and $\varphi \in BUC$, then $[\hat{M}_{\varphi_t}, A^{-1}] = A^{-1}[A, \hat{M}_{\varphi_t}]A^{-1} \rightrightarrows 0$ as $t \to 0$, which shows that also $A^{-1} \in BDO^p$. $\qquad\square$

We finally extend these notions to the continuous case.

Definition 1.47. *An operator A on L^p is called a* band operator*, a* band-dominated operator *or an operator in the* Wiener algebra *if its discretization* (1.3) *is so.*

1.3.7 Comparison

For simplicity of notation, again abbreviate the space $\ell^p(\mathbb{Z}^n, \mathbf{X})$ by E. We will finish this section with an illustrative and comprehensive comparison of the recently introduced operator classes

$$BO, \ BDO^p, \ K(E, \mathcal{P}) \ \text{ and } \ L(E, \mathcal{P}).$$

We will compare these classes from quite different points of view, which might turn out to be helpful for a better understanding of the different concepts.

Inclusions

Probably the first question that arises when comparing these classes is whether and how they are contained in each other. We already addressed this question for the classes $K(E)$, $K(E, \mathcal{P})$, $L(E)$ and $L(E, \mathcal{P})$ in Section 1.3.4. $K(E)$ and $L(E)$ shall not be considered here.
Obviously,

$$BO \subset \mathcal{W} \subset BDO^p \qquad \text{and} \qquad K(E, \mathcal{P}) \subset L(E, \mathcal{P}).$$

We will see how these two chains are connected.

Proposition 1.48. *The inclusion* $K(E, \mathcal{P}) \subset BDO^p \subset L(E, \mathcal{P})$ *holds.*

Proof. Let $T \in K(E, \mathcal{P})$, take an $\varepsilon > 0$, and choose $m \in \mathbb{N}$ such that $\|Q_m T\| < \varepsilon$ and $\|TQ_m\| < \varepsilon$. Then, for all sets $U, V \subset \mathbb{Z}^n$ with dist $(U, V) > d := 2m$, by

$$2m < \text{dist} (U, V) \leq \text{dist} (\{0\}, U) + \text{dist} (\{0\}, V),$$

at least one of the two terms on the right is greater than m. Suppose, without loss of generality, it is the second one. Then $P_V = P_V Q_m$, and consequently,

$$\|P_V T P_U\| \leq \|P_V Q_m T P_U\| \leq \|Q_m T\| < \varepsilon,$$

showing that $T \in \text{BDO}^p$ by Theorem 1.42.

Now suppose $A \in \text{BDO}^p$ and $k, m \in \mathbb{N}$ with $k \leq m$. With $U_m = \mathbb{Z}^n \setminus \{-m, \dots, m\}^n$ and $V_k = \{-k, \dots, k\}^n$, we have

$$P_k A Q_m = P_{V_k} A P_{U_m} \rightrightarrows 0 \qquad \text{and} \qquad Q_m A P_k = P_{U_m} A P_{V_k} \rightrightarrows 0$$

as dist $(U_m, V_k) = m - k + 1 \to \infty$ by Theorem 1.42. Consequently, for every fixed $k \in \mathbb{N}$, $P_k A Q_m \rightrightarrows 0$ and $Q_m A P_k \rightrightarrows 0$ as $m \to \infty$, and hence $A \in L(E, \mathcal{P})$ by Proposition 1.20. $\qquad\square$

As a consequence of Proposition 1.48, we get the following Venn diagram:

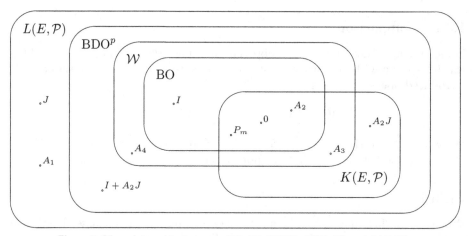

Figure 2: The relationship between BO, $\mathcal{W}$, BDO^p, $K(E, \mathcal{P})$ and $L(E, \mathcal{P})$.

The operators in the respective difference sets are as follows. Later in this section we will discuss some of these examples in more detail.

Example 1.49. a) We start with a first example in $L(E, \mathcal{P}) \setminus \text{BDO}^p$. Let J refer to the *flip operator* $(Ju)_\alpha = u_{-\alpha}$. Its matrix $[J]$ is supported on the cross diagonal only, all entries of which are $I_{\mathbf{X}}$.

b) Another example in $L(E, \mathcal{P}) \setminus \text{BDO}^p$ is $(A_1 u)_\alpha = u_{2\alpha}$. The matrix entries $a_{\alpha, 2\alpha}$ are equal to $I_{\mathbf{X}}$, and all other entries are zero.

c) A typical element of $K(E, \mathcal{P})$ is a generalized multiplication operator $A_2 = \hat{M}_b$ where the sequence $b = (b_\alpha)$ decays towards infinity. For example, we might put $b_\alpha = \frac{1}{|\alpha|+1} I_{\mathbf{X}}$, i.e. $A_2 = \hat{M}_f$ with $f(x) = \frac{1}{|x|+1}$.

d) The composition $A_2 J$ is an operator which is still in $K(E, \mathcal{P})$ (and consequently, in BDO^p) but no longer in BO, in contrast to A_2. The matrix $[A_2 J]$ is only supported on the cross diagonal where $a_{\alpha, -\alpha} = \frac{1}{|\alpha|+1} I_{\mathbf{X}}$.

e) Another example, very similar to $A_2 J$, is $(A_3 u)_\alpha = \frac{1}{2^{|\alpha|}} u_{-\alpha}$, the matrix representation of which also lives on the cross diagonal only, and $a_{\alpha, -\alpha} = \frac{1}{2^{|\alpha|}} I_{\mathbf{X}}$. This time the cross diagonal decays fast enough for $A \in \mathcal{W}$, in contrast to the previous example $A_2 J$.

f) For an example of a band-dominated operator A_4 which is neither in $K(E, \mathcal{P})$ nor a band operator, we pass to $n = 1$ and look at the Laurent operator (see Example 1.39)

$$A_4 = \sum_{k=-\infty}^{\infty} \frac{1}{2^{|k|}} V_k, \quad \text{i.e.} \quad [A_4] = \left[\frac{1}{2^{|i-j|}} I_{\mathbf{X}} \right]_{i,j \in \mathbb{Z}}.$$

Note that the choice of the coefficients $1/2^{|k|}$ implies $A_4 \in \mathcal{W}$ with $\|A_4\|_{\mathcal{W}} = 3$. $\quad \square$

Matrix Features

Here we will point out which properties of the matrix $[A]$ correspond to the membership of A in one of the classes BO, BDO^p, $K(E, \mathcal{P})$ and $L(E, \mathcal{P})$. Purely for illustration, and without any further explanation, we will insert some very intuitive pictures representing $[A]$.

- BO is the most trivial class from this point of view. By definition of this class, the matrix representation of $A \in$ BO is a band matrix – it is only supported on a finite number of diagonals. There is some number w, the band-width of A, such that all matrix entries $a_{\alpha\beta}$ with $|\alpha - \beta| > w$ are zero.

- BDO^p was derived from the previous class by passing to norm limits. For every $A \in \mathrm{BDO}^p$, the matrix $[A]$ now can be supported on infinitely many diagonals – but the entries in the γ-th diagonal have to tend to zero as $|\gamma| \to \infty$. We can easily derive this from Theorem 1.42, property (ii):

For every $\varepsilon > 0$, there is a $d > 0$ such that, for all entries $a_{\alpha\beta}$ with $|\alpha - \beta| > d$,

$$\|a_{\alpha\beta}\| = \|P_{\{\alpha\}} A P_{\{\beta\}}\| < \varepsilon.$$

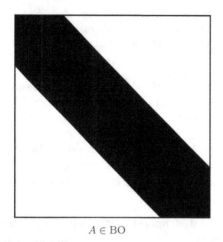

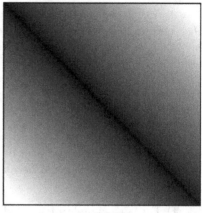

$A \in \mathrm{BO}$ $A \in \mathrm{BDO}^p$

Note that $A \in \mathrm{BDO}^p$ does not imply a particular decay rate of the norms $\|a_{\alpha\beta}\|$ as $|\alpha - \beta| \to \infty$. In fact, for every positive null sequence $\delta_0, \delta_1, \dots$ there is an operator $A \in \mathrm{BDO}^p$ such that

$$\sup_{|\alpha-\beta|=k} \|a_{\alpha\beta}\| = \delta_k$$

for all $k \in \mathbb{N}_0$. For example, take $(Au)_\alpha = \delta_{|\alpha|} u_{-\alpha}$. Then $[A]$ is only supported on the cross diagonal where $a_{\alpha,-\alpha} = \delta_{|\alpha|} I_{\mathbf{X}} \to 0$ as $|\alpha| \to \infty$.

On the other hand, an absolutely summable decay; that is

$$\delta_k := \sup_{|\alpha-\beta|=k} \|a_{\alpha\beta}\| \qquad \text{with} \qquad \sum_{k=0}^{\infty} \delta_k < \infty,$$

together with $A \in M(E)$, implies $A \in \mathcal{W} \subset \mathrm{BDO}^p$ for all $p \in [1, \infty]$.

- $K(E, \mathcal{P})$ is the set of all operators $A \in L(E)$ for which

$$Q_m A \rightrightarrows 0 \qquad \text{and} \qquad A Q_m \rightrightarrows 0 \qquad \text{as} \qquad m \to \infty.$$

The first condition implies that the matrix entries in the α-th row must tend to zero as $|\alpha| \to \infty$. The second condition implies that the same has to be true for columns. More precisely, for every $\varepsilon > 0$ there is a $m \in \mathbb{N}$ such that $\|a_{\alpha\beta}\| < \varepsilon$ for all matrix entries $a_{\alpha\beta}$ with $|\alpha| > m$ or $|\beta| > m$; that is, for all but finitely many entries. In short,

$$a_{\alpha\beta} \rightrightarrows 0 \qquad \text{as} \qquad (\alpha, \beta) \to \infty \text{ in } \mathbb{Z}^{2n}.$$

The finite submatrices $[a_{\alpha\beta}]_{|\alpha|,|\beta|\le m}$ already contain the "major part" of $[A]$, which is nicely reflected by the property $P_m A P_m \rightrightarrows A$ as $m \to \infty$.

- $L(E, \mathcal{P})$ is the set of all operators $A \in L(E)$ for which (1.12) holds. This is equivalent to

$$P_{\{\alpha\}} A Q_m \rightrightarrows 0 \qquad \text{and} \qquad Q_m A P_{\{\beta\}} \rightrightarrows 0 \qquad \text{as} \qquad m \to \infty$$

for all fixed $\alpha, \beta \in \mathbb{Z}^n$.

The first condition says that every row of the matrix $[A]$ decays at infinity. The second condition says that the same is true for every column.

$A \in K(E, \mathcal{P})$ $\qquad\qquad\qquad\qquad\qquad$ $A \in L(E, \mathcal{P})$

With this knowledge we have a look back to the operators in Figure 2 on page 28, see Example 1.49 a)–f).

The matrices $[J]$ and $[A_1]$ have entries of norm 1 in an arbitrarily large distance from the main diagonal, hence, these operators are not band-dominated. But every fixed row and every fixed column of $[J]$ and $[A_1]$ decays at infinity, so $J, A_1 \in L(E, \mathcal{P})$.

P_m, 0, A_2 and I are generalized multiplication operators. Consequently, they are band operators of band-width 0. Their matrix representations are supported on the main diagonal only. Those cases where this main diagonal decays at infinity, namely P_m, 0 and A_2, are in $K(E, \mathcal{P})$.

$A_2 J$ and A_3 are in $K(E, \mathcal{P})$ as well. But they are not band operators since their matrix entries are spread over all diagonals. From Proposition 1.48 we know that $A_2 J$ and A_3 are band-dominated (for all choices of $p \in [1, \infty]$). Note that the different decay rate of the entries of $[A_2 J]$ and $[A_3]$ towards infinity implies that A_3 is and $A_2 J$ is not contained in the Wiener-Algebra $\mathcal{W}$.

The matrix representation of $I + A_2 J$ is supported on the main and cross diagonal. The operator is not in $K(E, \mathcal{P})$ since the main diagonal does not vanish at infinity, but it is in BDO^p since the cross diagonal decays. However, this decay is not strong enough for membership in $\mathcal{W}$.

Finally, the Laurent operator A_4 is band-dominated. It is even in the Wiener algebra since $\sum_{k \in \mathbb{Z}} 1/2^{|k|} < \infty$. A_4 is not a band operator since it is supported on all diagonals, and it is not in $K(E, \mathcal{P})$ since its entries $a_{\alpha\beta}$ do not decay as $(\alpha, \beta) \to \infty$.

The operators in Example 1.26 a) and b) are not in $L(E, \mathcal{P})$ since the 0-th row does not decay at infinity. Example 1.26 c) is not in $M(E)$, whence looking at its matrix representation is completely misleading. In this case, the matrix leads us to the operator $0 \in M(E)$, which clearly differs from the operator in Example 1.26 c).

Information Transport

Finally, we will look at our operator classes from a very different point of view, which might be named "information transport" – at least we will do so. Again let E stand for $\ell^p(\mathbb{Z}^n, \mathbf{X})$ with $p \in [1, \infty]$ and $n \in \mathbb{N}$.

In (1.10), (1.11), (1.12), Proposition 1.36 and Theorem 1.42, the operator classes $K(E, \mathcal{P})$, $L(E, \mathcal{P})$, BO and BDO^p are characterized by the behaviour of $\|P_V A P_U\|$ for certain sets $U, V \subset \mathbb{Z}^n$ – for instance, when U and V are drifting apart. For a moment we will refer to U as "the source domain", to V as "the test domain", and to non-zero function values as "information".

Then $\|P_V A P_U\|$ measures the ability of the operator A to "transport" information from the source to the test domain. Indeed, starting with an arbitrary element $u \in E$ with $\|u\| = 1$, the information under consideration is restricted to the source domain by applying P_U to u. Then A maps $P_U u$ to an element $v := A P_U u$, and $\|P_V v\| = \|P_V A P_U u\|$ checks how much information has arrived at the test domain V.

Especially interesting is whether or not an operator A is able to transport information over large distances. Therefore, for every non-negative integer d, we define

$$f_A(d) := \sup_{U,V} \|P_V A P_U\|, \qquad (1.30)$$

where the supremum is taken over all $U, V \subset \mathbb{Z}^n$ with $\mathrm{dist}\,(U, V) \geq d$. The number $f_A(d)$ indicates A's best information transport ability, shortly called *information flow*, over the distance d.

Proposition 1.50. *For $A \in L(E)$, the function $f_A : \mathbb{N}_0 \to \mathbb{R}$ is monotonically decreasing with $f_A(0) = \|A\|$.*

Proof. $\{(U, V) : \mathrm{dist}\,(U, V) \geq d_1\} \supset \{(U, V) : \mathrm{dist}\,(U, V) \geq d_2\}$ shows that $f_A(d_1) \geq f_A(d_2)$ if $0 \leq d_1 < d_2$. With $U = V = \mathbb{Z}^n$ we have $d = 0$ and $f_A(0) \geq \|P_V A P_U\| = \|A\|$. On the other hand, for all sets $U, V \in \mathbb{Z}^n$, $\|P_V A P_U\| \leq \|P_V\| \cdot \|A\| \cdot \|P_U\| = \|A\|$, which shows that $f_A(0) \leq \|A\|$. $\qquad\qquad\square$

Since the function f_A is monotonically decreasing and bounded from below by zero, the limit

$$f_A(\infty) := \lim_{d \to \infty} f_A(d) \in \left[0, \|A\|\right]$$

always exists.

Remark 1.51. In the continuous case $E = L^p$, one might wish to consider $d \in [0, \infty)$ and $U, V \subset \mathbb{R}^n$ in (1.30), and consequently, $f_A : [0, \infty) \to \mathbb{R}$. This can lead to a more detailed picture about the behaviour of A, although the results of this approach are compatible with those stated here. □

From (1.16) we get that for every matrix entry of an operator $A \in L(E)$,

$$\|a_{\alpha\beta}\| \leq f_A(|\alpha - \beta|) \qquad \text{for all} \qquad \alpha, \beta \in \mathbb{Z}^n \tag{1.31}$$

holds. So $f_A(|\gamma|)$ is an upper bound on the matrix entries in the γ-th diagonal of A. Of course the function f_A cannot reflect the details of the matrix $[A]$ but, on the other hand, it detects some things that the matrix itself cannot 'see' – namely, if, roughly speaking, something is going on at infinity only. For instance, if C is as in Example 1.26 c), then $f_C(d) = \|C\|$ for all $d \in \mathbb{N}_0$.

Proposition 1.52. *If $A \in L(E)$ and $[A] = 0$, then $f_A \equiv \|A\|$.*

Proof. Let $\varepsilon > 0$ be arbitrary, and take $u \in E$ with $\|u\| = 1$ and $\|Au\| > \|A\| - \varepsilon/2$. From (1.7) we get that there exists a $k \in \mathbb{N}$ such that

$$\|P_k Au\| > \|Au\| - \varepsilon/2 > \|A\| - \varepsilon.$$

Since $[A] = 0$, we know from Proposition 1.31 a) that $AP_m = 0$, i.e $AQ_m = A$ for all $m \in \mathbb{N}$. Combining these two results, we get $\|P_k AQ_m u\| > \|A\| - \varepsilon$ for all $m \in \mathbb{N}$, showing that

$$f_A(d) \geq \|P_k AQ_{k+d}\| \geq \|P_k AQ_{k+d} u\| > \|A\| - \varepsilon$$

for all $d \geq 0$. Since ε was chosen arbitrarily and $f_A(d) \leq \|A\|$ for all $d \geq 0$, we are done. □

Now Proposition 1.36 and Theorem 1.42 allow the following characterizations of the classes BO and BDO^p in terms of the function f_A.

- $A \in$ BO with band-width w if and only if $f_A(w + 1) = 0$.

- In particular, A is a generalized multiplicator if and only if $f_A(1) = 0$.

- $A \in$ BDO^p if and only if $f_A(\infty) = 0$.

In words: Band operators are exactly those operators that cannot transport information over distances greater than a certain number (their band-width), while band-dominated operators are those operators that increasingly fail to transport information over a distance d as $d \to \infty$.

Example 1.53. As a very simple example, we look at the band Laurent operator $A = 3V_1 + 2V_5 + V_9$. We start with some information u supported in $\{-1, 0, 1\}$ and show the result Au as well as the information flow function f_A.

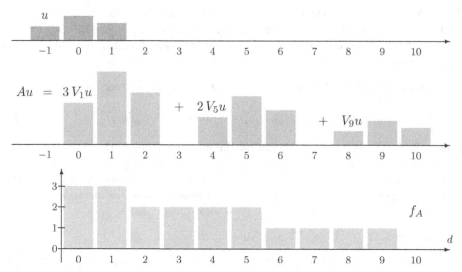

The matrix $[A]$ clearly has all entries equal to 3 on the 1st diagonal, equal to 2 on the 5th diagonal and equal to 1 on the 9th diagonal. The values and jumps of the function f_A correspond to these numbers. The band-width of A is 9.   □

Note that, in general, as the following example shows, it is not possible to find out whether an operator belongs to the Wiener algebra or not by looking at its information flow only.

Example 1.54. Suppose $n = 1$, and consider the operators $A, B \in \mathrm{BDO}^p$, the matrices of which are only supported on the cross diagonal, where

$$a_{k,-k} \;=\; \begin{cases} \frac{1}{2^m}, & k = 2^m, \\ 0 & \text{otherwise,} \end{cases} \qquad \text{and}$$

$$b_{k,-k} \;=\; \frac{1}{2^m} \quad \text{where} \quad 2^{m-1} < k \le 2^m$$

for $k \in \mathbb{N}$, and all other entries are zero. Then

$$\sum_{k=1}^{\infty} |a_{k,-k}| = 2, \qquad \text{while} \qquad \sum_{k=1}^{\infty} |b_{k,-k}| = \infty.$$

So, $A \in \mathcal{W}$ and $B \in \mathrm{BDO}^p \backslash \mathcal{W}$, although $f_A(d) = f_B(d) = b_{d,-d}$ for all $d \in \mathbb{N}_0$.   □

So there can be no condition on f_A which is sufficient and necessary for $A \in \mathcal{W}$. But at least we have the following result.

Proposition 1.55. *For $A \in L(E)$, the condition*

$$\sum_{d=1}^{\infty} d^{n-1} f_A(d) \;<\; \infty \tag{1.32}$$

is sufficient, but not necessary, for $A \in \mathcal{W}$.

Proof. Suppose $A \in L(E)$ and (1.32) holds. Then, clearly, $f_A(d) \to 0$ as $d \to \infty$, whence $A \in \mathrm{BDO}^p \subset M(E)$. So A is uniquely determined by its matrix representation $[A]$, and it remains to show that $\sum_{\gamma \in \mathbb{Z}^n} a_\gamma < \infty$, where a_γ is the supremum norm of the γ-th diagonal of $[A]$. From (1.31), we get that $a_\gamma \leq f_A(|\gamma|)$ for all $\gamma \in \mathbb{Z}^n$. Moreover, put $m_d := \#\{\gamma \in \mathbb{Z}^n : |\gamma| = d\}$ for every $d \in \mathbb{N}_0$. A little thought shows that

$$m_d = (2d+1)^n - (2d-1)^n \leq 2(2d+1)^{n-1} \leq 2(3d)^{n-1} = c\, d^{n-1}$$

holds for every $d \in \mathbb{N}$, where $c = 2 \cdot 3^{n-1}$ is independent of d. Consequently,

$$\sum_{\gamma \in \mathbb{Z}^n} a_\gamma \leq \sum_{\gamma \in \mathbb{Z}^n} f_A(|\gamma|) = \sum_{d=0}^{\infty} m_d\, f_A(d) \leq f_A(0) + c \sum_{d=1}^{\infty} d^{n-1} f_A(d) < \infty.$$

Finally, Example 1.54 shows that (1.32) is not necessary for $A \in \mathcal{W}$. □

Looking at (1.10), (1.11) and (1.12), we also find rough descriptions of the classes $L(E, \mathcal{P})$ and $K(E, \mathcal{P})$, although these are not as precisely as the characterizations of BO and BDO^p:

- $L(E, \mathcal{P})$ is the set of operators which do not draw information from any bounded set to infinity and no information from infinity to any bounded set.

- $K(E, \mathcal{P})$ is the set of operators which do not draw information from or to infinity at all.

All operators in Example 1.26 obviously draw information from infinity to some finite place. This already indicates that they are not in $L(E, \mathcal{P})$.

Example 1.56. a) If A is a generalized multiplication operator, $A = \hat{M}_b$, then clearly, $\|P_V A P_U\| = \|b|_{U \cap V}\|_\infty$ which is zero as soon as $\mathrm{dist}\,(U, V) > 0$. Multiplication operators are unable to transport information over any distance greater than zero. Their action is local – in the strictest sense of the word.

$$\begin{aligned} f_A(0) &= \|A\| = \|b\|_\infty \\ f_A(d) &= 0 \quad \text{if} \quad d \geq 1 \end{aligned}$$

b) For the shift operator $A = V_\alpha$ with $\alpha \in \mathbb{Z}^n$, we have $\|P_V A P_U\| = \|P_V P_{\alpha+U} V_\alpha\| = \|P_{(\alpha+U) \cap V}\|$. Consequently,

$$f_A(d) = \begin{cases} 1 & \text{if } 0 \leq d \leq |\alpha|, \\ 0 & \text{if } d > |\alpha|. \end{cases}$$

Of course, this answer is not very surprising: The shift operator V_α is only able to transport information over distances $d \leq |\alpha|$.

c) The flip operator $A = J$ from Example 1.49 a) is able to transport information over arbitrarily large distances (choose $U = -V$). The same is true for the operator in Example 1.49 b) (choose $U = \{2\alpha\}$ and $V = \{\alpha\}$). In both cases, $f_A \equiv 1$. Note that both operators draw information from infinity to infinity again (in the opposite direction, in the case of J) but not to any other place. So both operators are in $L(E, \mathcal{P})$. They are not in $\mathrm{BDO}^{\mathcal{P}}$ since $f_A(\infty) = 1 \neq 0$.   □

1.4   Invertibility of Sets of Operators

Fix a Banach space $\mathbf{X}$, and let T denote an arbitrary index set. Besides the invertibility of a single operator A on $\mathbf{X}$, we will be concerned with the question whether a whole set $\{A_\tau\}_{\tau \in T}$ of such operators is collectively invertible in a certain sense. Distinction is made between the following qualities of collective invertibility:

Definition 1.57. *We say that a bounded set $\{A_\tau\}_{\tau \in T}$ of operators $A_\tau \in L(\mathbf{X})$ is elementwise invertible if*

$$A_\tau \text{ is invertible for every } \tau \in T, \tag{E}$$

and we will call $\{A_\tau\}_{\tau \in T}$ uniformly invertible if

$$A_\tau \text{ is invertible for every } \tau \in T \quad \text{and} \quad \sup_{\tau \in T} \|A_\tau^{-1}\| < \infty. \tag{U}$$

Moreover, if $T \subset \mathbb{R}$ is measurable and $f : T \to L(\mathbf{X})$ denotes the mapping $\tau \mapsto A_\tau$, we will call $\{A_\tau\}_{\tau \in T}$ essentially invertible if

$$f \text{ is an invertible element of the Banach algebra } L^\infty(T, L(\mathbf{X})). \tag{S}$$

A first application of (U) is the following property which will be of great importance in the theory of approximation methods:

Definition 1.58. *Let T be some subset of $\mathbb{R}$ which is unbounded towards plus infinity. A bounded operator sequence $(A_\tau)_{\tau \in T}$ is said to be stable if there is some $\tau_* \in T$ such that the set $\{A_\tau\}_{\tau \in T, \, \tau > \tau_*}$ is uniformly invertible.*

In contrast to (U), property (S) only says that almost all operators A_τ are invertible, and that

$$\operatorname*{ess\,sup}_{\tau \in T} \|A_\tau^{-1}\| < \infty.$$

So there is a subset $T' \subset T \subset \mathbb{R}$ of measure zero such that an appropriate change of all A_τ with $\tau \in T'$ makes the set $\{A_\tau\}_{\tau \in T}$ uniformly invertible.

In Section 2.4 we have to deal with the question under which conditions on the set $\{A_\tau\}_{\tau \in T}$ property (S) implies (U). The following condition will prove to do the job.

Definition 1.59. *We call the mapping* $f : T \subset \mathbb{R} \to L(\mathbf{X})$ *with* $f : \tau \mapsto A_\tau$ *sufficiently smooth if it has the following property:*
If $(\tau_k)_{k=1}^\infty$ *is a sequence in* T *with* $\tau_k \to \tau_0 \in T$ *as* $k \to \infty$ *and* $(A_{\tau_k})_{k=1}^\infty$ *is stable, then* A_{τ_0} *is invertible and* $\|A_{\tau_0}^{-1}\| \leq \sup\limits_{k > k_*} \|A_{\tau_k}^{-1}\|$.

A first indication of the strength of this property is given by the following lemma. For simplicity, we will put $\|A^{-1}\| := \infty$ if A is not invertible. Moreover, put

$$U_\varepsilon^T(\tau) := (\tau - \varepsilon, \tau + \varepsilon) \cap T$$

for every $T \subset \mathbb{R}$, $\tau \in T$ and $\varepsilon > 0$.

Lemma 1.60. *Suppose* $\tau_0 \in T$ *is an accumulation point of* T, *and* $\tau \mapsto A_\tau$ *is sufficiently smooth. Then the following is true:*
If A_{τ_0} *is not invertible, then, for every* $M > 0$, *there is a neighborhood* $U_\varepsilon^T(\tau_0)$ *such that* $\|A_u^{-1}\| > M$ *for all* $u \in U_\varepsilon^T(\tau_0)$.

Proof. Conversely, suppose there is some $M > 0$ such that in every neighbourhood $U_{1/k}^T(\tau_0)$ with $k \in \mathbb{N}$, there is a τ_k with $\|A_{\tau_k}^{-1}\| \leq M$. Clearly, these τ_k form a sequence in T which converges to τ_0 as $k \to \infty$, and since $\tau \mapsto A_\tau$ is sufficiently smooth and (A_{τ_k}) is obviously stable, we get that A_{τ_0} is invertible, which is a contradiction. $\qquad\square$

Lemma 1.61. *Continuous functions* $f : T \to L(\mathbf{X})$ *are sufficiently smooth.*

Proof. Suppose $(\tau_k)_{k=1}^\infty \subset T$ tends to $\tau_0 \in T$ as $k \to \infty$ and A_{τ_k} is invertible for all sufficiently large k, say $k \geq k_*$, with

$$s := \sup_{k \geq k_*} \|A_{\tau_k}^{-1}\| < \infty.$$

If f is continuous, then $A_{\tau_k} \rightrightarrows A_{\tau_0}$ as $k \to \infty$, which is norm-convergence in the Banach algebra $L(\mathbf{X})$. So we just have to apply Lemma 1.3 to the sequence $(A_{\tau_k})_{k=k_*}^\infty$, to get that A_{τ_0} is invertible and $\|A_{\tau_0}^{-1}\| \leq s$. $\qquad\square$

Continuous functions are the most trivial examples of sufficiently smooth functions. In Chapter 2.4 we will study some more sophisticated cases, where $\tau \mapsto A_\tau$ is sufficiently smooth under considerably weaker conditions.

We will say that $T \subset \mathbb{R}$ is a *massive set* if $|U_\varepsilon^T(\tau)| > 0$ for every $\tau \in T$ and every $\varepsilon > 0$. Note that every interval (of positive length) is massive and that massive sets have no isolated points. $\mathbb{R} \setminus \mathbb{Q}$ is massive, $\mathbb{Q}$ is not.

Proposition 1.62. *Denote by* $f : T \to L(\mathbf{X})$ *the mapping* $\tau \mapsto A_\tau$. *Then the interplay between the properties* (E), (U) *and* (S) *of the set* $\{A_\tau\}_{\tau \in T}$ *is as follows:*

a) (U) *implies* (E).

b) *If* f *is continuous and* T *is compact, then* (E) *implies* (U).

In addition, suppose that T *is a measurable subset of* $\mathbb{R}$. *Then:*

c) (U) *implies* (S).

d) *If* f *is sufficiently smooth and* T *is massive, then* (S) *implies* (U).

Proof. a) and c) are obvious.

b) If $f : \tau \mapsto A_\tau$ is continuous and (E) is fulfilled, then also the mapping $\tau \mapsto \|A_\tau^{-1}\|$ is continuous. If, in addition, T is compact, then the latter mapping is bounded, and this implies (U).

d) Let (S) be fulfilled. Put

$$s := \operatorname*{ess\,sup}_{\tau \in T} \|A_\tau^{-1}\| \qquad \text{and} \qquad T' := \{\tau \in T : \|A_\tau^{-1}\| > s\}.$$

Suppose $\tau_0 \in T'$. Since T is massive, we have $|U_{1/k}^T(\tau_0)| > 0$ for every $k \in \mathbb{N}$. Since, by definition of ess sup, $|T'| = 0$, there is a $\tau_k \in U_{1/k}^T(\tau_0) \setminus T'$ for every $k \in \mathbb{N}$. Then $(\tau_k)_{k=1}^\infty \subset T \setminus T'$ with $\tau_k \to \tau_0$ as $k \to \infty$, and $(A_{\tau_k})_{k=1}^\infty$ is stable, by the choice of T'. But since f is sufficiently smooth, also A_{τ_0} is invertible with $\|A_{\tau_0}^{-1}\| \le s$, which contradicts $\tau_0 \in T'$. Consequently $T' = \varnothing$, and (U) holds.   □

1.5   Approximation Methods

HAGEN, ROCH and SILBERMANN begin [36] with the following neat overview:

- **Functional Analysis:** Solve equations in infinitely many variables.
- **Linear Algebra:** Solve equations in finitely many variables.
- **Numerical Analysis:** Build the bridge!

The latter is done by approximation methods.

1.5.1   Definition

Fix a Banach space E. If $A \in L(E)$ is an invertible operator, then the equation

$$Au = b \tag{1.33}$$

has a unique solution $u \in E$ for every right-hand side $b \in E$. We will deal with the approximate solution of this equation for $\dim E = \infty$ where $A \in L(E)$ and $b \in E$ are given and $u \in E$ is to be determined.

For this purpose, let $T \subset \mathbb{R}$ be some index set which is unbounded towards plus infinity, and let $(E_\tau)_{\tau \in T}$ refer to a sequence of Banach subspaces of E which are the images of projection operators $\Pi_\tau : E \to E_\tau$ and which exhausts E in the sense that the projections Π_τ converge[5] to the identity operator I on E as $\tau \to \infty$.

If one has to solve an equation of the form (1.33), one tries to approximate[5] the operator $A \in L(E)$ by a sequence of operators $(\tilde{A}_\tau)_{\tau \in T}$ with $\tilde{A}_\tau \in L(E_\tau)$ and to solve the somewhat simpler equations

$$\tilde{A}_\tau \tilde{u}_\tau \doteq \Pi_\tau b \qquad (1.34)$$

in E_τ instead, hoping these are uniquely solvable – at least for all sufficiently large τ – and that the solutions $\tilde{u}_\tau \in E_\tau$ of (1.34) tend[6] to the solution $u \in E$ of (1.33) as τ goes to infinity. If this is the case for every right-hand side $b \in E$, then we say that the *approximation method* $(\tilde{A}_\tau)$ is *applicable* to A.

This is the idea of approximation methods – or, to be more precisely, of projection methods. It is somewhat unsatisfactory that every operator of the sequence $(\tilde{A}_\tau)$ acts on a different space. To overcome these difficulties so that we may regard A and all operators of the approximation method as acting on the same space E, we will henceforth identify $\tilde{A}_\tau \in L(E_\tau)$ with $A_\tau := \tilde{A}_\tau + \Theta_\tau \in L(E)$, where $\Theta_\tau := I - \Pi_\tau$. Then, for every $\tau \in T$, (1.34) is equivalent to

$$A_\tau u_\tau = b \qquad \text{alias} \qquad \begin{pmatrix} \tilde{A}_\tau & 0 \\ 0 & \Theta_\tau \end{pmatrix} \begin{pmatrix} \tilde{u}_\tau \\ \Theta_\tau b \end{pmatrix} = \begin{pmatrix} \Pi_\tau b \\ \Theta_\tau b \end{pmatrix} \qquad (1.35)$$

with respect to the decomposition $E = E_\tau \dotplus \operatorname{im} \Theta_\tau$. The approximation method (1.34) is applicable to A if and only if (1.35) is so. Clearly, $\tilde{A}_\tau$ and A_τ are simultaneously invertible, where $\|A_\tau^{-1}\| = \max(\|\tilde{A}_\tau^{-1}\|, 1)$. Consequently, the sequence $(\tilde{A}_\tau)$ is stable (where every operator $\tilde{A}_\tau$ is acting on a different space E_τ) if and only if the sequence (A_τ) is.

A quite natural and very popular choice of the sequence (A_τ) is

$$\Pi_\tau A \Pi_\tau + \Theta_\tau, \qquad \tau \in T \qquad (1.36)$$

which we call the *natural projection method* for the operator A and the sequence (E_τ) of Banach spaces.

We will next make the index set T and the subspaces E_τ more precise for the spaces of our interest in this book, $E = \ell^p(\mathbb{Z}^n, \mathbf{X})$ and $E = L^p$.

1.5.2 Discrete Case

Let $E = \ell^p(\mathbb{Z}^n, \mathbf{X})$ with $p \in [1, \infty]$ and an arbitrary Banach space $\mathbf{X}$.

---

[5]The appropriate type of convergence heavily depends on the space E. A typical type for this purpose is the strong convergence. See Sections 1.5.5 and 1.6 for the type that we have in mind here.

[6]Again, the appropriate convergence type depends on E. This might be convergence in the norm of E or, as we have in mind, something else (also see Sections 1.5.5 and 1.6).

In the discrete case, there is usually no need for approximation methods with a continuous index set T. So the typical choice here shall be $T = \mathbb{N}$. For the definition of the subspaces E_τ with $\tau \in T = \mathbb{N}$, take a monotonously increasing sequence $(S_\tau)_{\tau \in \mathbb{N}}$ of finite subsets of $\mathbb{Z}^n$ which exhausts $\mathbb{Z}^n$ in the sense that, for every $\alpha \in \mathbb{Z}^n$, there is a $\tau_0 \in \mathbb{N}$ with $\alpha \in S_\tau$ for all $\tau \geq \tau_0$.

Then put $E_\tau := \ell^p(S_\tau, \mathbf{X})$, $\Pi_\tau = P_{S_\tau}$ and $\Theta_\tau = Q_{S_\tau}$. In this setting, we call (1.36) the *finite section method* for the discrete case and write $A_{\lceil \tau \rfloor}$ for the operator (1.36).

Example 1.63. Let $n = 1$, and put $S_\tau = \{-\tau, \ldots, \tau\}$ for every $\tau \in T = \mathbb{N}$. Then, in matrix language, the infinite system (1.33); that is

$$
\begin{pmatrix}
\ddots & \vdots & \vdots & \vdots & \ddots \\
\cdots & a_{-1,-1} & a_{-1,0} & a_{-1,1} & \cdots \\
\cdots & a_{0,-1} & a_{0,0} & a_{0,1} & \cdots \\
\cdots & a_{1,-1} & a_{1,0} & a_{1,1} & \cdots \\
\ddots & \vdots & \vdots & \vdots & \ddots
\end{pmatrix}
\begin{pmatrix}
\vdots \\
u(-1) \\
u(0) \\
u(1) \\
\vdots
\end{pmatrix}
=
\begin{pmatrix}
\vdots \\
b(-1) \\
b(0) \\
b(1) \\
\vdots
\end{pmatrix},
$$

is replaced by the sequence of finite systems (1.34), namely the truncations

$$
\begin{pmatrix}
a_{-\tau,-\tau} & \cdots & a_{-\tau,\tau} \\
\vdots & & \vdots \\
a_{\tau,-\tau} & \cdots & a_{\tau,\tau}
\end{pmatrix}
\begin{pmatrix}
\tilde{u}_\tau(-\tau) \\
\vdots \\
\tilde{u}_\tau(\tau)
\end{pmatrix}
=
\begin{pmatrix}
b(-\tau) \\
\vdots \\
b(\tau)
\end{pmatrix},
$$

for $\tau = 1, 2, \ldots$. This is where the name 'finite section method' comes from.

During this procedure one is keeping fingers crossed that the latter systems are uniquely solvable once they are sufficiently large, and that $\tilde{u}_\tau(k) \to u(k)$ as $\tau \to \infty$ for every $k \in \mathbb{Z}$. □

1.5.3 Continuous Case

Let $E = L^p = L^p(\mathbb{R}^n) = L^p(\mathbb{R}^n, \mathbb{C})$ with $p \in [1, \infty]$.

In the continuous case, it is often convenient to work with approximation methods with a continuous index set T. The typical choice here is $T = \mathbb{R}_+ = (0, \infty)$. For the definition of the subspaces E_τ with $\tau \in T = \mathbb{R}_+$, take a monotonously increasing sequence $(\Omega_\tau)_{\tau > 0}$ of compact subsets of $\mathbb{R}^n$ which exhausts $\mathbb{R}^n$ in the sense that for every $x \in \mathbb{R}^n$ there is a $\tau_0 > 0$ with $x \in \Omega_\tau$ for all $\tau \geq \tau_0$.

Then put $E_\tau := L^p(\Omega_\tau)$, $\Pi_\tau = P_{\Omega_\tau}$ and $\Theta_\tau = Q_{\Omega_\tau}$. In this setting, we call (1.36) the *finite section method* for the continuous case and write $A_{\lceil \tau \rfloor}$ for the operator (1.36).

1.5.4 Additional Approximation Methods

At the first glance, one might be moaning about the practicability of the finite section method in the continuous case. Clearly, in the discrete case, equation (1.33) is an infinite linear system of equations. This is reduced to a finite one (for instance, see Example 1.63) by passing to (1.34), where the latter is readily solved numerically – whereas, in the continuous case, the equations (1.34) are still too complicated to be solved directly.

What one usually does for the numerical solution of the equations (1.34) in $L^p(\Omega_\tau)$ is to "superimpose" another approximation method onto it in order to discretize the operators $\tilde{A}_\tau$. The choice of this method heavily depends on the space E and the operator A. For example, if $E = \mathrm{BC} \subset L^\infty$ (see Subsection 4.2.3), then a series of discretization methods for the finitely truncated equation are known. But this shall not be our topic here.

However, note that by passing from (1.33) to (1.34), a major step for the approximate solution of (1.33) is done. Therefore, we will restrict ourselves to the study of methods of the form (1.34), like the finite section method.

1.5.5 Which Type of Convergence is Appropriate?

Up to now we did not specify clearly in which way we will expect the operators A_τ (and $\tilde{A}_\tau$) and the solutions u_τ (and $\tilde{u}_\tau$) to approximate A and u, respectively. Therefore, let us fix a $p \in [1, \infty]$ and a Banach space $\mathbf{X}$, and let E denote one of the spaces $\ell^p(\mathbb{Z}^n, \mathbf{X})$ and L^p.

It is easily seen that, for $p = \infty$, it is not appropriate to expect strong convergence of A_τ to A – as one usually does if $p < \infty$ – since not even the finite section projectors $\Pi_\tau = P_{S_\tau}$ or $\Pi_\tau = P_{\Omega_\tau}$, which are the identity operators on E_τ, converge strongly to the identity I on E as $\tau \to \infty$.

The same argument shows that, for the finite section method, also the solutions u_τ will not converge to u in the norm of E if $p = \infty$ – as it is the case for $p < \infty$. We will introduce the appropriate type of convergence of A_τ to A and u_τ to u in Definition 1.64 in the following section.

1.6 $\mathcal{P}$-convergence

Let $T \subset \mathbb{R}$ be an index set which is unbounded towards plus infinity. If we say that the *final part* of a sequence $(A_\tau)_{\tau \in T}$ has a certain property, then we mean that there exists a $\tau_* \in T$ such that the subsequence $(A_\tau)_{\tau \in T, \tau > \tau_*}$ has this property. For example, a sequence is stable if and only if its final part is uniformly invertible. For convergence issues, clearly, the final part of a sequence is of interest only.

Now fix a $p \in [1, \infty]$ and a Banach space $\mathbf{X}$. In this section, E denotes one of the spaces $\ell^p(\mathbb{Z}^n, \mathbf{X})$ and L^p, and $\mathcal{P} = (P_1, P_2, \ldots)$ is the corresponding approximate identity.

In Section 1.3.4 we have seen that, if $1 < p < \infty$, then $K(E)$ determines *-strong convergence in terms of (1.8) and (1.9). In the same sense, $K(E, \mathcal{P})$ determines the convergence in which our approximate identity $\mathcal{P}$ approximates the identity operator I on E. But this is exactly the convergence that we need in Section 1.5.

1.6.1   Definition and Equivalent Characterization

$K(E, \mathcal{P})$ was introduced in order to characterize the convergence of P_m to I as $m \to \infty$. Now we will study this convergence type in detail.

Definition 1.64. *A sequence* $(A_\tau)_{\tau \in T} \subset L(E)$ *is* $\mathcal{P}$-*convergent to* $A \in L(E)$ *if*

$$\|K(A_\tau - A)\| \to 0 \qquad and \qquad \|(A_\tau - A)K\| \to 0 \qquad as \qquad \tau \to \infty \qquad (1.37)$$

for every $K \in K(E, \mathcal{P})$, *and a sequence* $(u_\tau)_{\tau \in T} \subset E$ *is called* $\mathcal{P}$-*convergent to* $u \in E$ *if*

$$\|K(u_\tau - u)\| \to 0 \qquad as \qquad \tau \to \infty \qquad (1.38)$$

for every $K \in K(E, \mathcal{P})$. *We will write* $A_\tau \overset{P}{\to} A$ *or* $A = \mathcal{P}\text{–}\lim A_\tau$ *for* (1.37), *and* $u_\tau \overset{P}{\to} u$ *or* $u = \mathcal{P}\text{–}\lim u_\tau$ *for* (1.38).

It is readily seen that a $\mathcal{P}$-limit is unique if it exists. Indeed, suppose for every operator $K \in K(E, \mathcal{P})$, we have $K(A_\tau - A) \rightrightarrows 0$ and $K(A_\tau - B) \rightrightarrows 0$; and consequently, $K(A - B) \rightrightarrows 0$ as $\tau \to \infty$, i.e. $K(A - B) = 0$. Then clearly, $P_m(A - B) = 0$ for all $m \in \mathbb{N}$, whence $A - B = 0$ by Lemma 1.30 a). Analogously, we see that also $\mathcal{P}\text{–}\lim u_\tau$ is unique.

If we want to check whether a sequence $(A_\tau) \subset L(E)$ or $(u_\tau) \subset E$ is $\mathcal{P}$-convergent, it is not necessary to check (1.37) or (1.38) for **all** $K \in K(E, \mathcal{P})$, as the following proposition shows.

Proposition 1.65.

a) *Let* $(A_\tau)_{\tau \in T} \subset L(E)$ *and* $A \in L(E)$ *be arbitrary. Then* $A_\tau \overset{P}{\to} A$ *if and only if the final part of the sequence* (A_τ) *is bounded, and*

$$\|P_m(A_\tau - A)\| \to 0 \qquad and \qquad \|(A_\tau - A)P_m\| \to 0 \qquad as \qquad \tau \to \infty$$
$$(1.39)$$

for every fixed $m \in \mathbb{N}$.

b) *Let* $(u_\tau)_{\tau \in T} \subset E$ *and* $u \in E$ *be arbitrary. Then* $u_\tau \overset{P}{\to} u$ *if and only if the final part of the sequence* (u_τ) *is bounded, and*

$$\|P_m(u_\tau - u)\| \to 0 \qquad as \qquad \tau \to \infty \qquad (1.40)$$

for every fixed $m \in \mathbb{N}$.

Remark 1.66. a) If $\{\tau \in T : \tau \leq \tau_*\}$ is finite for every $\tau_* \in T$, then the boundedness of the final part of a sequence is equivalent to the boundedness of the whole sequence of course.

b) If we replace $L(E)$ by $L(E, \mathcal{P})$ in a), then we get a well-known fact, which is already proven in [70] and [50]. The new fact here is that (1.37) implies the boundedness condition on (A_τ) also for arbitrary operators in $L(E)$! (For operators in $L(E,\mathcal{P})$, this is a simple consequence of Proposition 1.69 and the Banach-Steinhaus theorem.)

Note that, for A and A_τ in $L(E)$, already the first property in (1.37) is sufficient for the boundedness of the final part of (A_τ). The second property in (1.37) does not imply the boundedness of $(A_\tau)_{\tau \in T, \tau > \tau_*}$ for any $\tau_* > 0$, as we can see if we put $A_\tau = \tau C$ where C is the operator from Example 1.26 c).

c) Part b) of Proposition 1.65 shows that $\mathcal{P}$-convergence of sequences $(u_\tau) \subset E$ is equivalent to strict convergence in the sense of BUCK [12]. $\square$

Proof. a) Suppose $(A_\tau)_{\tau \in T, \tau > \tau_*}$ is bounded for some $\tau_* > 0$ and (1.39) holds. Then, for all $m \in \mathbb{N}$ and all $K \in K(E,\mathcal{P})$, one has

$$\|K(A_\tau - A)\| \ \leq \ \|K\|\,\|P_m(A_\tau - A)\| + \|KQ_m\|\,\|A_\tau - A\|,$$

where the first term tends to zero as $\tau \to \infty$, and the second one is as small as desired if m is large enough. The second property of (1.37) is shown absolutely analogously.

Conversely, if (1.37) holds for all $K \in K(E,\mathcal{P})$, then (1.39) holds for all $m \in \mathbb{N}$ since $\mathcal{P} \subset K(E,\mathcal{P})$. It remains to show that $(A_\tau)_{\tau \in T, \tau > \tau_*}$ is bounded for some $\tau_* > 0$.

Suppose the converse is true. Without loss of generality, we can suppose that $A = 0$. Now we will successively define two sequences: $(m_k)_{k=1}^\infty \subset \mathbb{N}$ and $(\tau_k)_{k=0}^\infty \subset T \cup \{0\}$. We start with $m_1 := 1$ and $\tau_0 := 0$.

For every $k \in \mathbb{N}$, choose $\tau_k \in T$ such that

$$\tau_k > \tau_{k-1}\,, \qquad \|A_{\tau_k}\| > k^2 + 3 \qquad \text{and} \qquad \|P_{m_k}A_{\tau_k}\| < 1,$$

the latter possible since $P_{m_k} \in K(E,\mathcal{P})$ and $A_\tau \xrightarrow{\mathcal{P}} 0$. Then

$$\|Q_{m_k}A_{\tau_k}\| \ \geq \ \|A_{\tau_k}\| - \|P_{m_k}A_{\tau_k}\| \ > \ k^2 + 3 - 1 = k^2 + 2.$$

Take $u_k \in E$ with $\|u_k\| = 1$ and $\|Q_{m_k}A_{\tau_k}u_k\| > k^2 + 1$, and choose $m_{k+1} > m_k$ such that $\|P_{m_{k+1}}Q_{m_k}A_{\tau_k}u_k\| > k^2$, which is possible by (1.7). Consequently,

$$\|P_{m_{k+1}}Q_{m_k}A_{\tau_k}\| > k^2 \qquad \text{for all} \qquad k \in \mathbb{N}.$$

Now consider the generalized multiplication operator

$$K := \sum_{j=1}^\infty \frac{1}{j^2}\, P_{m_{j+1}}Q_{m_j}.$$

Then it is easily seen that $K \in K(E,\mathcal{P})$. But on the other hand, from

$$P_{m_{k+1}}Q_{m_k}P_{m_{j+1}}Q_{m_j} = \begin{cases} P_{m_{k+1}}Q_{m_k}, & j = k, \\ 0, & j \neq k, \end{cases}$$

we get that

$$\|KA_{\tau_k}\| \geq \|P_{m_{k+1}}Q_{m_k}KA_{\tau_k}\| = \|\frac{1}{k^2}P_{m_{k+1}}Q_{m_k}A_{\tau_k}\| > \frac{k^2}{k^2} = 1$$

for every $k \in \mathbb{N}$, which contradicts $\|KA_\tau\| \to 0$ as $\tau \to \infty$.

b) can be verified in an analogous way. $\square$

Corollary 1.67. *For bounded sequences $(A_\tau) \subset L(E)$, the $\mathcal{P}$-convergence to $A \in L(E)$ is equivalent to (1.37) for all $K \in \mathcal{P}$.*

The notion of $\mathcal{P}$-convergence also yields a nice alternative characterization of the class $L(E, \mathcal{P})$: An operator $A \in L(E)$ is in $L(E, \mathcal{P})$ if and only if both sequences

$$(AQ_m)_{m=1}^\infty \quad \text{and} \quad (Q_m A)_{m=1}^\infty$$

$\mathcal{P}$-converge to zero as $m \to \infty$. If this is even norm-convergence, then $A \in K(E, \mathcal{P})$.

We will see that the study of $\mathcal{P}$-convergence in $L(E, \mathcal{P})$ leads to especially nice results, which might be seen as indication that $L(E, \mathcal{P})$ is the natural playground for $\mathcal{P}$-convergence.

1.6.2 $\mathcal{P}$-convergence in $L(E, \mathcal{P})$

Proposition 1.68. *$L(E, \mathcal{P})$ is sequentially closed with respect to $\mathcal{P}$-convergence.*

Proof. Take $(A_\tau) \subset L(E, \mathcal{P})$ with $A_\tau \xrightarrow{\mathcal{P}} A$. We will show that $A \in L(E, \mathcal{P})$ as well. Fix a $k \in \mathbb{N}$ and let $\varepsilon > 0$. Take $\tau_0 \in T$ large enough that $\|P_k(A_{\tau_0}-A)\| < \varepsilon/2$, and choose $m_0 \in \mathbb{N}$ large enough that $\|P_k A_{\tau_0} Q_m\| < \varepsilon/2$ for all $m > m_0$. Then

$$\|P_k AQ_m\| \leq \|P_k(A_{\tau_0} - A)\|\,\|Q_m\| + \|P_k A_{\tau_0} Q_m\| < \varepsilon/2 + \varepsilon/2 = \varepsilon$$

for all $m > m_0$ which shows that $P_k AQ_m \rightrightarrows 0$ as $m \to \infty$. The symmetric property $Q_m AP_k \rightrightarrows 0$ as $m \to \infty$ can be verified analogously. $\square$

Since $K(E, \mathcal{P})$ is an ideal in $L(E, \mathcal{P})$, one can associate with every $A \in L(E, \mathcal{P})$ the two operators of left and right multiplication, acting **on** $K(E, \mathcal{P})$ by

$$A^l : K \mapsto AK \quad \text{and} \quad A^r : K \mapsto KA$$

for every $K \in K(E, \mathcal{P})$. This observation and the following two propositions are due to Roch, Silbermann and Rabinovich [70].

Proposition 1.69.   a) $\|A^l\|_{L(K(E,\mathcal{P}))} = \|A\| = \|A^r\|_{L(K(E,\mathcal{P}))}$ *for all $A \in L(E, \mathcal{P})$.*

b) *A sequence $(A_\tau) \subset L(E, \mathcal{P})$ is $\mathcal{P}$-convergent to some $A \in L(E, \mathcal{P})$ if and only if $A_\tau^l \to A^l$ and $A_\tau^r \to A^r$ strongly in $L(K(E, \mathcal{P}))$.*

Proof. a) Take an arbitrary $A \in L(E, \mathcal{P})$. Clearly, $\|A^l\| \leq \|A\|$, and $\|A^r\| \leq \|A\|$.

For the reverse inequalities, take an arbitrary $\varepsilon > 0$, and choose $u \in E$ with $\|u\| = 1$ such that $\|Au\| \geq \|A\| - \varepsilon$. Moreover, choose $k \in \mathbb{N}$ with $\|Au\| - \|P_k Au\| \leq \varepsilon$, which is possible by (1.7). Then $\|P_k Au\| \geq \|A\| - 2\varepsilon$. Since $P_k \in K(E, \mathcal{P})$, we have

$$\|A^r\| = \|A^r\| \|P_k\| \geq \|P_k A\| \geq \|P_k Au\| \geq \|A\| - 2\varepsilon. \tag{1.41}$$

Choosing $m \in \mathbb{N}$ large enough that $\|P_k A Q_m\| \leq \varepsilon$ and taking the inequality $\|P_k A\| \geq \|A\| - 2\varepsilon$ from (1.41), we conclude from $P_m \in K(E, \mathcal{P})$ to

$$\|A^l\| = \|A^l\| \|P_m\| \geq \|A P_m\| \geq \|P_k A P_m\| \geq \|P_k A\| - \|P_k A Q_m\| \geq \|A\| - 3\varepsilon. \tag{1.42}$$

Since (1.41) and (1.42) hold for every $\varepsilon > 0$, we are done with a).

b) By a), $L(E, \mathcal{P})$ is isometrically embedded into $L(K(E, \mathcal{P}))$ by each one of the mappings $A \mapsto A^l$ and $A \mapsto A^r$. Consequently, $\|K(A_\tau - A)\| \to 0$ for all $K \in K(E, \mathcal{P})$ is equivalent to the strong convergence of A_τ^r to A^r on $K(E, \mathcal{P})$, and the second property in (1.37) is the same for A_τ^l and A^l. $\qquad \square$

Proposition 1.70. *For sequences $(A_\tau), (B_\tau) \subset L(E, \mathcal{P})$ with $\mathcal{P}$-limits A and B, respectively, we have*

$$
\begin{array}{lll}
\text{a)} & \|A\| \leq \liminf \|A_\tau\| \leq \sup \|A_\tau\| < \infty, \\
\text{b)} & \mathcal{P}\text{-}\lim(A_\tau + B_\tau) = A + B, \\
\text{c)} & \mathcal{P}\text{-}\lim(A_\tau B_\tau) = AB.
\end{array}
$$

Proof. From Proposition 1.68 we get that $A, B \in L(E, \mathcal{P})$. Using Proposition 1.69, we only have to recall the Banach-Steinhaus theorem as well as the compatibility of strong limits with addition and composition.

Alternatively, there are elementary proofs for b) and c) using Proposition 1.65. Equality b) is trivial anyway, and c) can be seen as follows. For all $k, m \in \mathbb{N}$,

$$
\begin{aligned}
\|P_k(A_\tau B_\tau - AB)\| &\leq \|P_k(A_\tau - A)B_\tau\| + \|P_k A(B_\tau - B)\| \\
&\leq \|P_k(A_\tau - A)\| \|B_\tau\| \\
&\quad + \|P_k A\| \|P_m(B_\tau - B)\| + \|P_k A Q_m\| \|B_\tau - B\|
\end{aligned}
$$

holds, where the first two terms tend to zero as $\tau \to \infty$, and the third one becomes as small as desired if we choose m large enough (note that, by Proposition 1.65, the $\mathcal{P}$-convergent sequence (B_τ) is bounded). $(A_\tau B_\tau - AB)P_k \rightrightarrows 0$ is shown completely symmetric using the boundedness of (A_τ). $\qquad \square$

Remark 1.71. The previous two propositions show that $L(E, \mathcal{P})$ is indeed a very nice playground for the study of $\mathcal{P}$-convergence. Originally we even conjectured that $A_\tau \xrightarrow{\mathcal{P}} A$ implies that $A_\tau - A = B_\tau + C_\tau$ with a sequence $(B_\tau) \subset L(E, \mathcal{P})$ and $C_\tau \rightrightarrows 0$, which would make the study of $\mathcal{P}$-convergence outside of $L(E, \mathcal{P})$ completely ridiculous. But the following example reveals that this is not true. $\quad \square$

Example 1.72. Take $E = \ell^\infty(\mathbb{Z})$ and $T = \mathbb{N}$ as index set. Then the sequence $(A_k)_{k=1}^\infty$ with

$$A_k\, u = Q_k\,(\ldots, u_k, u_k, u_k, \ldots)$$

for all $u \in E$, is bounded and clearly subject to (1.39) with $A = 0$, whence it is $\mathcal{P}$-convergent to zero. On the other hand, the k-th column of A_k has all but finitely many entries equal to 1. This shows that $A_k \notin L(E, \mathcal{P})$ – more precisely, $\mathrm{dist}\,(A_k, L(E, \mathcal{P})) = 1$ for all $k \in \mathbb{N}$. So obviously, $A_m - A = A_m$ is more than a null sequence C_m away from all $B_m \in L(E, \mathcal{P})$.   □

1.6.3 $\mathcal{P}$-convergence vs. $*$-strong Convergence

From the discussion at the beginning of Section 1.6 we see that the relation between $K(E)$ and $K(E, \mathcal{P})$ essentially determines the relation between $\mathcal{P}$- and $*$-strong convergence. In analogy to Figure 1 on page 15, we will briefly study this relation here, depending on the space $E = \ell^p(\mathbb{Z}^n, \mathbf{X})$.

From (1.8), (1.9) and Definition 1.64 we get that $A_\tau \xrightarrow{P} A$ implies $A_\tau \xrightarrow{K} A$ if $K(E) \subset K(E, \mathcal{P})$, i.e. if $1 < p < \infty$, and that $A_\tau \xrightarrow{K} A$ implies $A_\tau \xrightarrow{P} A$ if $K(E, \mathcal{P}) \subset K(E)$, i.e. if $\dim \mathbf{X} < \infty$. Moreover, by Corollary 1.16, $A_\tau \xrightarrow{*} A$ always implies $A_\tau \xrightarrow{K} A$, and both are equivalent if E is reflexive; that is if $\mathbf{X}$ is reflexive (recall that every finite-dimensional space $\mathbf{X}$ is reflexive) and $1 < p < \infty$. Consequently, we get that

$$A_\tau \xrightarrow{P} A \implies A_\tau \xrightarrow{K} A \iff A_\tau \xrightarrow{*} A \qquad \text{if } 1 < p < \infty \text{ and } \mathbf{X} \text{ reflexive,}$$
$$A_\tau \xrightarrow{*} A \implies A_\tau \xrightarrow{K} A \implies A_\tau \xrightarrow{P} A \qquad \text{if } \dim \mathbf{X} < \infty.$$

The following table shows the sign that the star $\circledast$ can be substituted with, depending on p and $\mathbf{X}$. The implication '$\Leftarrow$' in the right column of the second row holds under the assumption that $\mathbf{X}$ is reflexive, for example $\mathbf{X} = L^p(H)$.

			$\circledast$	$\dim \mathbf{X} < \infty$	$\dim \mathbf{X} = \infty$
			$p = 1$	$\implies$	
$A_\tau \xrightarrow{*} A$	$\circledast$	$A_\tau \xrightarrow{P} A$	$1 < p < \infty$	$\iff$	$\Longleftarrow$
			$p = \infty$	$\implies$	

Figure 3: $\mathcal{P}$-convergence versus $*$-strong convergence, depending on the space $E = \ell^p(\mathbb{Z}^n, \mathbf{X})$.

An example, indicating why '$\Rightarrow$' is not true in the right column of the second row, is the sequence $(A_\tau) = (P_{[0,\frac{1}{\tau}]})_{\tau=1}^\infty$ on L^p with $1 < p < \infty$.

For $p = 1$ and $p = \infty$, we have $P_m \xrightarrow{P} I$ and $P_m \xrightarrow{*}\!\!\!\!\!/\ \ I$ as $m \to \infty$, showing that '$\Leftarrow$' is not true in the first and third row of the table. But for $p = 1$, however, 50% of $P_m \xrightarrow{*} I$ is true since we have strong convergence $P_m \to I$ here. For $p = \infty$, only $P_m^\triangleleft \to I^\triangleleft$ is true, provided $\mathbf{X}^\triangleleft$, and hence $E^\triangleleft$ exists. The following two propositions show that this behaviour is a rule.

Proposition 1.73. If $p = 1$, then $A_\tau \xrightarrow{P} A$ implies the strong convergence $A_\tau \to A$.

Proof. If $A_\tau \xrightarrow{\mathcal{P}} A$, then we know from Proposition 1.65 that the final part of (A_τ) is bounded and that $(A_\tau - A)P_m \rightrightarrows 0$ as $\tau \to \infty$ for every $m \in \mathbb{N}$. Consequently, for every $K \in K(E)$ and every $m \in \mathbb{N}$,

$$\|(A_\tau - A)K\| \leq \|(A_\tau - A)P_m\|\,\|K\| + \|A_\tau - A\|\,\|Q_mK\|,$$

where the first term tends to zero as $\tau \to \infty$, and the second term can be made as small as desired if we choose m large enough since K is compact and $Q_m \to 0$ strongly as $m \to \infty$. By Proposition 1.13, this implies the strong convergence of A_τ to A as $\tau \to \infty$. $\qquad\square$

Proposition 1.74. *If $p = \infty$ and $\mathbf{X}^\lhd$ as well as the pre-adjoint operators $A_\tau^\lhd$, $A^\lhd$ exist, then $A_\tau \xrightarrow{\mathcal{P}} A$ implies the strong convergence $A_\tau^\lhd \to A^\lhd$ on $E^\lhd$.*

Proof. If $E = \ell^\infty(\mathbb{Z}^n, \mathbf{X})$ and $\mathbf{X}^\lhd$ exists, then $E^\lhd = \ell^1(\mathbb{Z}^n, \mathbf{X}^\lhd)$. Now proceed as in the proof of Proposition 1.73 to get

$$\|(A_\tau^\lhd - A^\lhd)K\| \leq \|(A_\tau^\lhd - A^\lhd)P_m^\lhd\|\,\|K\| + \|A_\tau^\lhd - A^\lhd\|\,\|Q_m^\lhd K\|,$$

for all $K \in K(E^\lhd)$. Now $\|(A_\tau^\lhd - A^\lhd)P_m^\lhd\| = \|P_m(A_\tau - A)\| \to 0$ as $\tau \to \infty$, and $\|Q_m^\lhd K\|$ can be made arbitrarily small by choosing $m \in \mathbb{N}$ large enough since $Q_m^\lhd \to 0$ on $E^\lhd$. Again, application of Proposition 1.13 finally proves the claim. $\quad\square$

1.7 Applicability vs. Stability

Let $E = \ell^p(\mathbb{Z}^n, \mathbf{X})$ or $E = L^p$ with $p \in [1, \infty]$, and let $T \subset \mathbb{R}$ be a index set which is unbounded towards plus infinity. We will see that the question whether an approximation method $(A_\tau)_{\tau \in T}$ is applicable or not, heavily depends on the stability (cf. Definition 1.58) of the sequence (A_τ).

The classic Polski theorem [36] says that a **strongly** convergent approximation method (A_τ) is applicable to A – with convergence of the solutions u_τ to u in the **norm** of E – if and only if the operator A is invertible and the sequence (A_τ) is stable. It was proven by ROCH and SILBERMANN in [76] (also see Theorem 6.1.3 in [70]) that the same is true for the applicability of the methods that we have in mind: $\mathcal{P}$-**convergent** methods (A_τ) with $\mathcal{P}$-**convergent** solutions u_τ – provided that the sequence (A_τ) is subject to the following condition:

We write $(A_\tau) \in \mathcal{F}(E, \mathcal{P})$ if the sequence (A_τ) is bounded, and, for every $k \in \mathbb{N}$,

$$\sup_{\tau \in T} \|P_k A_\tau Q_m\| \to 0 \quad \text{and} \quad \sup_{\tau \in T} \|Q_m A_\tau P_k\| \to 0 \quad \text{as} \quad m \to \infty. \quad (1.43)$$

Of course, therefore it is necessary that $A_\tau \in L(E, \mathcal{P})$ for every $\tau \in T$, and moreover $A \in L(E, \mathcal{P})$ by Proposition 1.68 if $A_\tau \xrightarrow{\mathcal{P}} A$.

Theorem 1.75. *If $(A_\tau) \in \mathcal{F}(E, \mathcal{P})$ is $\mathcal{P}$-convergent to A, then the approximation method (A_τ) is applicable to A if and only if A is invertible and (A_τ) is stable.*

Proof. See Theorem 6.1.3 of [70] or [76] or 4.41f in [58].   □

In the case of a discrete index set T, the uniformity condition (1.43) is redundant if $(A_\tau)_{\tau \in T} \subset L(E, \mathcal{P})$. For simplicity, we state and prove this result for $T = \mathbb{N}$.

Lemma 1.76. *If $(A_\tau)_{\tau \in \mathbb{N}} \subset L(E, \mathcal{P})$ is $\mathcal{P}$-convergent to A, then $(A_\tau) \in \mathcal{F}(E, \mathcal{P})$.*

Proof. The boundedness of the sequence follows from Proposition 1.65. It remains to prove (1.43). Therefore fix an arbitrary $k \in \mathbb{N}$. From $(A_\tau) \subset L(E, \mathcal{P})$ we conclude that for every $\varepsilon > 0$ and every $\tau \in \mathbb{N}$ there is a $m(\tau, \varepsilon) \in \mathbb{N}$ such that $\|P_k A_\tau Q_m\| < \varepsilon$ for all $m > m(\tau, \varepsilon)$. What we have to show is that

$$\forall \varepsilon > 0 \quad \exists m(\varepsilon) : \qquad \|P_k A_\tau Q_m\| < \varepsilon \qquad \forall m > m(\varepsilon) \quad \forall \tau \in \mathbb{N}. \qquad (1.44)$$

So take an arbitrary $\varepsilon > 0$ and choose $m_0 \in \mathbb{N}$ large enough that $\|P_k A Q_m\| < \varepsilon/2$ for all $m > m_0$, which is possible since $A \in L(E, \mathcal{P})$ by Proposition 1.68. Moreover, choose $\tau_0 \in \mathbb{N}$ large enough that $\|P_k (A_\tau - A)\| < \varepsilon/2$ for all $\tau > \tau_0$.

For $m > m_0$ and $\tau > \tau_0$ we conclude

$$\|P_k A_\tau Q_m\| \leq \|P_k A Q_m\| + \|P_k (A_\tau - A)\| \, \|Q_m\| < \varepsilon/2 + \varepsilon/2 = \varepsilon.$$

Now choose $m(\varepsilon) := \max(\, m(1, \varepsilon), \, \dots, \, m(\tau_0, \varepsilon), \, m_0 \,)$ to ensure (1.44). Analogously, we prove the second property in (1.43).   □

From Theorem 1.75 and Lemma 1.76 we immediately get the following.

Corollary 1.77. *If $(A_\tau)_{\tau \in \mathbb{N}} \subset L(E, \mathcal{P})$ is $\mathcal{P}$-convergent to A, then the approximation method (A_τ) is applicable to A if and only if A is invertible and (A_τ) is stable.*

Also for arbitrary index sets T, the property $(A_\tau) \in \mathcal{F}(E, \mathcal{P})$ is often automatically implied by the nature of the approximation method. We will check this for the finite section method in $E = L^p$.

Proposition 1.78. *The finite section method $(A_{\lceil \tau \rceil})_{\tau \in T}$ is applicable to $A \in L(E, \mathcal{P})$ if and only if A is invertible and the finite section sequence $(A_{\lceil \tau \rceil})$ is stable.*

Proof. Let $(\Omega_\tau)_{\tau \in T}$ denote the increasing and exhausting (in the sense of Subsection 1.5.3) sequence of compact subsets of $\mathbb{R}^n$. Then the sequence $(A_{\lceil \tau \rceil})$, defined by

$$A_{\lceil \tau \rceil} = P_{\Omega_\tau} A P_{\Omega_\tau} + Q_{\Omega_\tau},$$

is automatically contained in $L(E, \mathcal{P})$ if $A \in L(E, \mathcal{P})$, it $\mathcal{P}$-converges to A as $\tau \to \infty$ by Proposition 1.70, and it is obviously bounded with $\|A_{\lceil \tau \rceil}\| \leq \|A\| + 1$ for all $\tau \in T$.

By Theorem 1.75, it remains to prove (1.43). Therefore, take an arbitrary $k \in \mathbb{N}$, and note that, for all $\tau \in T$ and $m \in \mathbb{N}$,

$$
\begin{aligned}
\|P_k A_{\lceil \tau \rceil} Q_m\| &\leq \|P_k P_{\Omega_\tau} A P_{\Omega_\tau} Q_m\| + \|P_k Q_{\Omega_\tau} Q_m\| \\
&= \|P_{\Omega_\tau} P_k A Q_m P_{\Omega_\tau}\| + \|P_k Q_m Q_{\Omega_\tau}\| \\
&\leq \|P_k A Q_m\| + \|P_k Q_m\|,
\end{aligned}
$$

which shows that $\sup_{\tau \in T} \|P_k A_{\lceil \tau \rceil} Q_m\| \leq \|P_k A Q_m\| + \|P_k Q_m\| \to 0$ as $m \to \infty$ since $A \in L(E, \mathcal{P})$ and $P_k Q_m = 0$ if $m \geq k$. Analogously, one checks the second property in (1.43). □

1.8 Comments and References

The study of concrete classes of band and band-dominated operators (such as convolution, Wiener-Hopf, and Toeplitz operators) goes back to the 1950's starting with [40] and [30] by GOHBERG and KREIN and was culminating in the 1970/80's with the huge monographs [29] by GOHBERG/FELDMAN and [9] by BÖTTCHER and SILBERMANN. The study of band-dominated operators as a general operator class was initiated by SIMONENKO [82], [83]. More recent work along the lines of this book can be found in [43], [67], [58] and [41], to mention some examples only. In [58] band-dominated operators are called "operators of local type" and in [41] "operators with uniformly fading memory". The history of Theorem 1.42 goes back to [43], [67], [50] and [70].

Most of the theory presented in this chapter has been done before in [41], [36], [50] and [70], for example. Some results in this chapter are probably new. The most interesting of these might be Propositions 1.24 (with consequences for Figure 1, of course), 1.65 and Lemma 1.76.

The idea to study ℓ^p-sequences with values in a Banach space **X** and to identify L^p with such a space is not new, of course. It can be found in [41], [8], [36], [68] and [70], for example.

Approximate identities are introduced as special approximate projections in [70]. Its applications go far beyond the finite section projectors in ℓ^p and L^p presented here. For more examples see [36], [70] and [17].

Generalized compactness notions were introduced in [8], [76], [36], [68] and [70], for example. We adapt the main ideas and the notations $K(E, \mathcal{P})$ and $L(E, \mathcal{P})$ from [70]. Operators in $L(E, \mathcal{P})$ are called "operators with locally fading memory" in [41]. Proposition 1.24, which is an important ingredient to the clarification of the relations between $K(E)$ and $K(E, \mathcal{P})$, goes back to CHANDLER-WILDE and the author [17].

$\mathcal{P}$-convergence of operator sequences was studied in [76], [58], [68] and [70], for example. Theorem 1.75 goes back to [76], [58] and [70].

Chapter 2

Invertibility at Infinity

In this chapter we discuss the property that will play a central role throughout this book: *invertibility at infinity*.

2.1 Fredholm Operators

Let E denote an arbitrary Banach space. By a theorem of BANACH, $A \in L(E)$ is invertible if and only if it is injective and surjective as a mapping $A : E \to E$. Thus, if A is not invertible, then $\ker A \neq \{0\}$ or $\operatorname{im} A \neq E$, or both. As an indication how badly injectivity and surjectivity of A are violated, one defines the two numbers

$$\alpha := \dim \ker A \qquad \text{and} \qquad \beta := \dim \operatorname{coker} A, \qquad (2.1)$$

where $\operatorname{coker} A := E/\operatorname{im} A$, provided that the image of A is closed.

Recall that $A \in L(E)$ is referred to as a *Fredholm operator* if its image is closed and both numbers α and β are finite. In that case, their difference

$$\operatorname{ind} A := \alpha - \beta$$

is called the *index* of A.

It turns out that $A \in L(E)$ is a Fredholm operator if and only if there are operators $B, C \in L(E)$ such that

$$AB = I + T_1 \qquad \text{and} \qquad CA = I + T_2 \qquad (2.2)$$

hold with some $T_1, T_2 \in K(E)$. The operators B and C are called right and left Fredholm regularizers of A, respectively.

By evaluating the term CAB, it becomes evident that B and C only differ by an operator in $K(E)$. Consequently, B (as well as C) is a regularizer from both sides, showing that A is Fredholm if and only if there is an operator $B \in L(E)$ such that

$$AB = I + T_1 \qquad \text{and} \qquad BA = I + T_2 \qquad (2.3)$$

hold with some $T_1, T_2 \in K(E)$.

Obviously, (2.3) is equivalent to the invertibility of the coset $A + K(E)$ in the factor algebra $L(E)/K(E)$ where $(A + K(E))^{-1} = B + K(E)$. In a slightly more fluent language, we will reflect this by saying that A is invertible modulo $K(E)$ and B is an inverse of A modulo $K(E)$.

Although this is very well known, we will emphasize this algebraic characterization of the Fredholm property as a proposition because it underlines an important connection between operator theory and the theory of Banach algebras. For a more detailed coverage of the theory of Fredholm operators, including proofs, see e.g. [31].

Proposition 2.1. $A \in L(E)$ is a Fredholm operator if and only if $A + K(E)$ is invertible in the factor algebra $L(E)/K(E)$.

In analogy to the spectrum of an operator as an element of $L(E)$, we introduce the Fredholm spectrum or the so-called *essential spectrum* of $A \in L(E)$ as

$$\mathrm{sp_{ess}}\, A = \{\lambda \in \mathbb{C} \, : \, A - \lambda I \text{ is not Fredholm}\}.$$

By Proposition 2.1 we have that $\mathrm{sp_{ess}}\, A = \mathrm{sp}_{L(E)/K(E)}(A + K(E))$.

The factor algebra $L(E)/K(E)$ is the so-called *Calkin algebra*. It is well-known that, if a coset $A + K(E)$ is invertible in the Calkin algebra, then all elements of $A + K(E)$ are Fredholm operators with the same index. For instance, all operators in $I + K(E)$ are Fredholm and have index zero.

The theory of Fredholm operators is a rigorous generalization of the probably best known statement connected with the name of ERIK IVAR FREDHOLM: *Fredholm's alternative.*

In terms of (2.1), it says that, for an operator $A = I + K$ with $K \in K(E)$,

either $\alpha = 0$ and $\beta = 0$,   or alternatively, $\alpha \neq 0$ and $\beta \neq 0$.

This statement is clearly true for all Fredholm operators $A \in L(E)$ with $\alpha = \beta$; that is $\{A \in L(E) \, : \, A \text{ Fredholm, ind}\, A = 0\}$. This class is strictly larger than $I + K(E)$, for which Fredholm's alternative is usually formulated. It contains all operators of the form $A = C + K$ where $C \in L(E)$ is invertible and $K \in K(E)$. A simple argument (see [31, Theorem 6.2]) shows that it actually coincides with this class.

2.2 Invertibility at Infinity

Fix a Banach space **X** and let $p \in [1, \infty]$. As in Chapter 1, we abbreviate the Banach space $\ell^p(\mathbb{Z}^n, \mathbf{X})$ by E.

Recall that we substituted the pair $\big(L(E) \,; K(E) \big)$ by $\big(L(E, \mathcal{P}) \,; K(E, \mathcal{P}) \big)$ in Section 1.3.4. In analogy to Fredholmness of $A \in L(E)$, which corresponds to an invertibility problem in $L(E)/K(E)$, this naturally leads to the study of the property of an operator $A \in L(E, \mathcal{P})$ that corresponds to the invertibility of $A + K(E, \mathcal{P})$ in $L(E, \mathcal{P})/K(E, \mathcal{P})$. We will prove that this invertibility is equivalent to the existence of operators $B, C \in L(E, \mathcal{P})$ and an integer $m \in \mathbb{N}$ such that

$$Q_m AB = Q_m = CAQ_m \qquad (2.4)$$

holds.

Proposition 2.2. *For all $A \in L(E, \mathcal{P})$, the following properties are equivalent:*

(i) *The coset $A + K(E, \mathcal{P})$ is invertible in $L(E, \mathcal{P})/K(E, \mathcal{P})$.*

(ii) *There exist $B, C \in L(E, \mathcal{P})$ with (2.2) for some $T_1, T_2 \in K(E, \mathcal{P})$.*

(iii) *There exists a $B \in L(E, \mathcal{P})$ with (2.3) for some $T_1, T_2 \in K(E, \mathcal{P})$.*

(iv) *There exist $B, C \in L(E, \mathcal{P})$ and an $m \in \mathbb{N}$ such that (2.4) holds.*

Proof. Obviously, (i) is equivalent to both (ii) and (iii).

(iii)$\Rightarrow$(iv). Take $m \in \mathbb{N}$ large enough that $\|Q_m T_1\| < 1$ and $\|T_2 Q_m\| < 1$, and put $D := (I + Q_m T_1)^{-1}$. From the Neumann series (or less elementary, from Proposition 1.46), we get that $D \in L(E, \mathcal{P})$. Then, by (2.3) and $Q_m = Q_m^2$,

$$Q_m AB = Q_m + Q_m T_1 = Q_m(I + Q_m T_1),$$

and consequently, $Q_m AB' = Q_m$ with $B' = BD \in L(E, \mathcal{P})$. The second equality in (2.4) follows from a symmetric argument using $BA = I + T_2$.

(iv)$\Rightarrow$(iii). If (iv) holds, then

$$AB = Q_m AB + P_m AB = Q_m + P_m AB = I - P_m + P_m AB =: I + T,$$

where $T = P_m AB - P_m \in K(E, \mathcal{P})$ since $A, B \in L(E, \mathcal{P})$. The second claim in (2.3) follows analogously. $\square$

Remark 2.3. The operators B and C from (ii) only differ by an operator in $K(E, \mathcal{P})$. Consequently, both can be used as the operator B in (iii). Also note that m in (2.4) clearly can be replaced by any $m' > m$. $\square$

In accordance with [70], we will call $A \in L(E, \mathcal{P})$ a $\mathcal{P}$-*Fredholm operator* if property (i) holds. Since this notion is not defined for operators outside of $L(E, \mathcal{P})$, we will also study a very similar property in the somewhat larger setting $A \in L(E)$ which enables us to compare the new property with usual Fredholmness on equal territory:

Definition 2.4. *An operator* $A \in L(E)$ *is called invertible at infinity if there exists an operator* $B \in L(E)$ *such that (2.3) holds for some* $T_1, T_2 \in K(E, \mathcal{P})$. *Moreover,* $A \in L(E)$ *is called weakly invertible at infinity if there exist* $B, C \in L(E)$ *and an* $m \in \mathbb{N}$ *such that (2.4) holds.*

Remark 2.5. a) Of course, A is invertible at infinity if it is invertible, as we see by putting $B = A^{-1}$ and $T_1 = T_2 = 0$ in (2.3).

b) Note that, if $A \in L(E, \mathcal{P})$ is invertible at infinity, we do not know if it is even $\mathcal{P}$-Fredholm, since we cannot guarantee[1] that B from Definition 2.4 can be chosen from $L(E, \mathcal{P})$ if $A \in L(E, \mathcal{P})$. But the reverse implication is true of course: If $A \in L(E, \mathcal{P})$ is $\mathcal{P}$-Fredholm, then it is invertible at infinity. Moreover, we do have the following. □

Proposition 2.6. *If* $A \in L(E)$ *is invertible at infinity, then it is weakly invertible at infinity.*

Proof. The proof of Proposition 2.2 (iii)⇒(iv) works as well (it is even simpler) if we replace $L(E, \mathcal{P})$ by $L(E)$. □

We would not have chosen two different names if the two properties were equivalent. Indeed, here we give four examples of operators which are only weakly invertible at infinity.

Example 2.7. a) Let $p = 1$, $n = 1$, and let K denote the first operator in Example 1.26 a). Our example here is $A := I - K$ (note that $A := I + K$ is no good choice since this operator is even invertible with $A^{-1} = I - K/2$).

From $Q_m K = 0$ we conclude $Q_m A = Q_m$ for all $m \in \mathbb{N}_0$, and consequently, (2.4) with $B = C = Q_1$ and $m = 1$, which shows that A is weakly invertible at infinity. Concerning invertibility at infinity, note that the second equality in (2.3) holds with $B = Q_1$ and $T_2 = -P_1 \in K(E, \mathcal{P})$ for example. But we will show that the first equality in (2.3) cannot be fulfilled with $B \in L(E)$ and $T_1 \in K(E, \mathcal{P})$.

Therefore suppose that there exist such B and T_1 where $AB = I + T_1$ holds. From $Q_m B - Q_m = Q_m A B - Q_m = Q_m (AB - I) = Q_m T_1$ we conclude

$$Q_m B - Q_m \rightrightarrows 0 \qquad \text{as} \qquad m \to \infty.$$

Secondly, without loss of generality, we can suppose that $(Bu)_0 = 0$ for all $u \in E$ since A ignores the 0-th component of its operand. Consequently, $P_{\{0\}} B Q_m = 0$ for all $m \in \mathbb{N}$. Moreover, for all $i \in \mathbb{Z} \setminus \{0\}$,

$$P_{\{i\}} B Q_m = P_{\{i\}} A B Q_m + P_{\{i\}} K B Q_m \rightrightarrows 0 \qquad \text{as} \qquad m \to \infty$$

since $AB = I + T_1 \in L(E, \mathcal{P})$ and $P_{\{i\}} K = 0$ if $i \neq 0$. Choosing an arbitrary $k \in \mathbb{N}$ and summing up $P_{\{i\}} B Q_m$ over all $i = -k, \dots, k$, we get

$$P_k B Q_m \rightrightarrows 0 \qquad \text{as} \qquad m \to \infty.$$

---

[1] An appropriate modification of the proof of Theorem 1.1.9 in [70] was not successful – or not appropriate.

Now let $\varepsilon > 0$ and choose $k \in \mathbb{N}$ large enough that $D := Q_k - Q_k B$ has norm $\|D\| < \varepsilon/\|A\|$. Then for all $m \geq k$,

$$
\begin{aligned}
\|P_0 A Q_m\| &= \|P_0 A Q_k Q_m\| \leq \|P_0 A Q_k B Q_m\| + \|P_0 A D Q_m\| \\
&\leq \|P_0 A B Q_m\| + \|P_0 A P_k B Q_m\| + \|P_0 A D Q_m\| \\
&\leq \|P_0 A B Q_m\| + \|A\| \cdot \|P_k B Q_m\| + \|A\| \cdot \varepsilon/\|A\|,
\end{aligned}
$$

which shows that $\|P_0 A Q_m\| \to 0$ as $m \to \infty$ since $AB = I + T_1 \in L(E, \mathcal{P})$ and $\|P_k B Q_m\| \to 0$ as $m \to \infty$. Having a look at our operator A, we see that this is wrong as $\|P_0 A Q_m\| = 1$ for all $m \in \mathbb{N}$. Contradiction.

b) Again let $p = 1$ and $n = 1$. But now substitute K in a) by the second operator (denoted by $\tilde{A}$) of Example 1.26 a). The rest is analogously.

c) and d) Let $p = \infty$ and consider the adjoint operators A^* of a) and b), respectively. By duality arguments we see that also A^* is only weakly invertible at infinity. $\qquad\square$

From the definition of a $\mathcal{P}$-Fredholm operator it is immediately clear that this property is robust under perturbations in $K(E, \mathcal{P})$ and under perturbations with sufficiently small norm, provided the latter are in $L(E, \mathcal{P})$. While, for invertibility at infinity in $L(E)$, we conjecture that this is not true, at least the following can be shown.

Proposition 2.8. *Weak invertibility at infinity in $L(E)$ is robust*

   a) *under perturbations in $K(E, \mathcal{P})$, and*

   b) *under small perturbations in $L(E)$.*

Proof. Suppose $A \in L(E)$, and there exist $m \in \mathbb{N}$ and $B, C \in L(E)$ such that (2.4) holds.

a) Let $T \in K(E, \mathcal{P})$, and choose $m \in \mathbb{N}$ such that, in addition to (2.4), also $\|Q_m T\| < 1/\|B\|$ holds. Then from

$$
Q_m(A + T)B = Q_m AB + Q_m TB = Q_m + Q_m^2 TB = Q_m(I + Q_m TB)
$$

and the invertibility of $I + Q_m TB$, we get that $Q_m(A + T)B' = Q_m$ holds, where $B' = B(I + Q_m TB)^{-1}$. The second equality in (2.4) is checked in a symmetric way.

b) Let $S \in L(E)$ be subject to $\|S\| \leq 1/\|B\|$. Then from

$$
Q_m(A + S)B = Q_m AB + Q_m SB = Q_m + Q_m SB = Q_m(I + SB)
$$

and the invertibility of $I + SB$, we get $Q_m(A + S)B' = Q_m$ with $B' = B(I + SB)^{-1}$. Again, the second equality in (2.4) is checked in a symmetric way. $\qquad\square$

2.2.1   Invertibility at Infinity in BDO^p

Since we are especially interested in band-dominated operators, we insert some results on the invertibility at infinity of operators in BDO^p.

Proposition 2.9. $K(E,\mathcal{P})$ *is an ideal in* BDO^p.

Proof. This is trivial since $K(E,\mathcal{P}) \subset \mathrm{BDO}^p \subset L(E,\mathcal{P})$ by Proposition 1.48, and $K(E,\mathcal{P})$ is an ideal in $L(E,\mathcal{P})$ by Proposition 1.19.   □

Proposition 2.9 makes it possible to study the factor algebra $\mathrm{BDO}^p/K(E,\mathcal{P})$. We will now prove that, if $A \in \mathrm{BDO}^p$ is invertible at infinity, the operator B in (2.3) is automatically in BDO^p as well.

Proposition 2.10. *For* $A \in \mathrm{BDO}^p$, *the following properties are equivalent.*

  (i) *A is invertible at infinity.*

 (ii) $A + K(E,\mathcal{P})$ *is invertible in the factor algebra* $\mathrm{BDO}^p/K(E,\mathcal{P})$.

(iii) *A is $\mathcal{P}$-Fredholm.*

Proof. Basically, (i), (ii) and (iii) all say that there exists an operator B such that (2.3) holds for some $T_1, T_2 \in K(E,\mathcal{P})$. The difference is, where B can be found: (i) says $B \in L(E)$, (ii) says $B \in \mathrm{BDO}^p$, and (iii) says $B \in L(E,\mathcal{P})$. So clearly, (ii)⇒(iii)⇒(i) holds.

    (i)⇒(ii). Suppose $B \in L(E)$, and (2.3) holds with $T_1, T_2 \in K(E,\mathcal{P})$.

    Take some $\varphi \in \mathrm{BUC}$, and define $\psi := \varphi - \varphi(0) \in \mathrm{BUC}$. Then, for every $t \in \mathbb{R}^n$,

$$[\hat{M}_{\varphi_t}, B] = [\hat{M}_{\psi_t}, B] = B[A, \hat{M}_{\psi_t}]B + C_t \tag{2.5}$$

holds, where φ_t and ψ_t are defined as in (1.22), and $C_t = (T_2 \hat{M}_{\psi_t})B - B(\hat{M}_{\psi_t} T_1)$, by (2.3). But for every $m \in \mathbb{N}$,

$$\|\hat{M}_{\psi_t} T_1\| \leq \|(\hat{M}_{\psi_t} P_m) T_1\| + \|\hat{M}_{\psi_t}(Q_m T_1)\|$$

holds, where the second term on the right can be made as small as desired by choosing m large enough, and the first term evidently tends to 0 as $t \to 0$ since $\psi(0) = 0$. The same argument for $\|T_2 \hat{M}_{\psi_t}\|$ shows that $C_t \rightrightarrows 0$ as $t \to 0$. Together with (2.5) and Theorem 1.42, we get $B \in \mathrm{BDO}^p$, and hence (ii).   □

Corollary 2.11. $\mathrm{BDO}^p/K(E,\mathcal{P})$ *is inverse closed in* $L(E,\mathcal{P})/K(E,\mathcal{P})$.

2.3   Invertibility at Infinity vs. Fredholmness

On page 15 we promised to come back to the diagrams in Figure 1 later. Now it is again time to do so.

    The aim of this section is to study the dependencies between invertibility at infinity and Fredholmness on $E = \ell^p(\mathbb{Z}^n, \mathbf{X})$ in the six essential cases determined by p and dim $\mathbf{X}$. We will derive a table showing which of the two properties implies the other in which of the six cases. That table is proven to be complete by giving appropriate counter-examples for the "missing" implications.

The relation of the sets $K(E)$ and $K(E, \mathcal{P})$ essentially determines the relation of the properties of Fredholmness and invertibility at infinity: Clearly, looking at (2.2), if $K(E) \subset K(E, \mathcal{P})$, then Fredholmness implies invertibility at infinity. Conversely, if $K(E, \mathcal{P}) \subset K(E)$, then invertibility at infinity implies Fredholmness.

Consequently, Figure 1 on page 15 immediately implies the following table which shows the sign that the star $\circledast$ can be substituted with, depending on p and $\mathbf{X}$:

invertibility at ∞ $\circledast$ Fredholmness		$\circledast$	dim $\mathbf{X} < \infty$	dim $\mathbf{X} = \infty$
		$p = 1$	$\Longrightarrow$	
		$1 < p < \infty$	$\Longleftrightarrow$	$\Longleftarrow$
		$p = \infty$	$\Longrightarrow$	

Figure 4: Invertibility at infinity versus Fredholmness, depending on the space $E = \ell^p(\mathbb{Z}^n, \mathbf{X})$.

As a justification for the missing implication arrows in this table, we will give some counter-examples. But before, there is one more addition to Figure 4: Although '$\Leftarrow$' does not hold in the upper left corner of our table, the following proposition shows that it "almost" holds, in the following sense.

Proposition 2.12. *Let* $p = 1$ *and* dim $\mathbf{X} < \infty$. *If* $A \in L(E)$ *is Fredholm, then it is weakly invertible at infinity.*

Proof. For the proof of the first '=' sign in (2.4), take $B \in L(E)$ and $T \in K(E)$ with $AB = I + T$. Since $Q_m \to 0$, we have $Q_m T \rightrightarrows 0$ as $m \to \infty$. Choose $m \in \mathbb{N}$ large enough that $\|Q_m T\| < 1$ and put $D := I + Q_m T$, which is invertible then. Now

$$Q_m A B = Q_m(I + T) = Q_m + Q_m T = Q_m(I + Q_m T) = Q_m D$$

shows that $Q_m A B' = Q_m$ with $B' = BD^{-1}$.

For the second '=' sign in (2.4), we first claim that, for all sufficiently large $m \in \mathbb{N}$,

$$\ker AQ_m = \ker Q_m \qquad (2.6)$$

holds. Clearly, '$\supset$' holds in (2.6). Suppose '$\subset$' does not hold. Then there is a sequence $(m_k)_{k=1}^\infty \subset \mathbb{N}$ tending to infinity such that, for every $k \in \mathbb{N}$, there is a $x_k \in \ker AQ_{m_k} \setminus \ker Q_{m_k}$, i.e.

$$y_k := Q_{m_k} x_k \neq 0 \qquad \text{where} \qquad Ay_k = AQ_{m_k} x_k = 0 \qquad \text{for all} \qquad k \in \mathbb{N}.$$

But these $y_k \in \operatorname{im} Q_{m_k}$ clearly span an infinite-dimensional space, which contradicts dim $\ker A < \infty$. Consequently, (2.6) holds.

Equality (2.6) with a sufficiently large $m \in \mathbb{N}$ shows that E decomposes into a direct sum of $\ker AQ_m = \ker Q_m = \operatorname{im} P_m$ and $\operatorname{im} Q_m$. From $E = \ker AQ_m \dotplus \operatorname{im} Q_m$ we conclude that

$$AQ_m|_{\operatorname{im} Q_m} : \operatorname{im} Q_m \to \operatorname{im} AQ_m \qquad (2.7)$$

is a bijection. Since A and Q_m are Fredholm (remember that $\dim \mathbf{X} < \infty$), we get that AQ_m is Fredholm, and hence, that $E_1 := \operatorname{im} AQ_m$ is closed and complementable. Now let E_2 denote a complementary space of E_1 and define an operator $C \in L(E)$ which acts on E_1 as the inverse of (2.7) and is zero on E_2. With that construction, $CAQ_m = Q_m$ holds.   □

To show that in the first and third row of our table, the implication '$\Leftarrow$' cannot hold in general, we give four examples of Fredholm operators which are not invertible (although weakly invertible) at infinity.

For this purpose we can reuse Example 2.7. If $\dim \mathbf{X} < \infty$, then all four operators, a), b), c) and d) are Fredholm (since K is compact) and weakly invertible at infinity – but not invertible at infinity. For $\dim \mathbf{X} = \infty$, examples a) and c) are no longer Fredholm, but b) and d) still have this property as K is still compact then.

For an operator which is Fredholm but not even weakly invertible at infinity, we go to $p = \infty$. We draw some inspiration from Example 1.26 c).

Example 2.13. Put $p = \infty$, let E_0 denote the space of all $u \in E$ which decay at infinity, and fix an element $v \in E \setminus E_0$. By the Hahn-Banach theorem there exists a bounded linear functional f on E which vanishes on E_0 and takes $f(v) = 1$. Then define operators K and A on E by

$$K : u \mapsto f(u) \cdot v \qquad \text{and} \qquad A := I - K.$$

Clearly, K is compact. Consequently, A is Fredholm with index zero. A simple computation shows that the kernel of A exactly consists of the multiples of v, whence $\alpha = 1$. Then also $\beta = 1$ follows. But how can we identify this one-dimensional cokernel?

The fact that

$$f(Au) = f\Big(u - f(u)v\Big) = f(u) - f(u)f(v) = 0 \qquad (2.8)$$

for every $u \in E$, is a strong indication that A is not surjective (and hence, not right-invertible) at infinity. Roughly speaking, the cokernel of A must be located somewhere at infinity.

Indeed, suppose that A were weakly (right-)invertible at infinity, i.e. there exist $B \in L(E)$ and $m \in \mathbb{N}$ such that $Q_m AB = Q_m$ holds. If we rewrite this equality as $AB - P_m AB = I - P_m$, let both sides act on v and apply the functional f, we arrive at the contradiction

$$0 - 0 = f(ABv) - f(P_m ABv) = f(v) - f(P_m v) = 1 - 0$$

by (2.8), by $\operatorname{im} P_m \subset E_0$ and by the choice of f.   □

Moreover, if dim $\mathbf{X} = \infty$, then Q_m is an example of an operator which is invertible at infinity but not Fredholm, where $m \in \mathbb{N}$ is arbitrary. Also note that a multiplication operator M_b on L^p is invertible at infinity if and only if its symbol $b \in L^\infty$ is essentially bounded away from zero in a neighborhood of infinity – while M_b is Fredholm if and only if this is true on the whole $\mathbb{R}^n$. This shows that in the right column of our table, the implication '$\Rightarrow$' cannot hold in general. This example finishes the completeness discussion for our table in Figure 4.

Later, in Section 4.2, we will encounter a class of operators for which we can guarantee the implication '$\Rightarrow$' also in the right column of the table:

Definition 2.14. *We say that an operator $W \in L(E)$ is locally compact if $P_U W$ and $W P_U$ are compact for all bounded sets $U \subset \mathbb{Z}^n$.*

Proposition 2.15. *Let $A = S + W \in L(E)$, where S is invertible and W is locally compact. If A is invertible at infinity, then A is Fredholm.*

Proof. Suppose there are $B \in L(E)$ and $T_1, T_2 \in K(E, \mathcal{P})$ such that (2.3) holds. Now put $R := (I - BW)S^{-1}$. Then

$$
\begin{aligned}
AR &= A(I - BW)S^{-1} = AS^{-1} - ABWS^{-1} = I + WS^{-1} - (I + T_1)WS^{-1} \\
&= I - T_1 WS^{-1}.
\end{aligned}
$$

From $T_1 W = T_1(P_m W) + (T_1 Q_m)W$ we see that $T_1 W$ is the norm limit of the compact operators $T_1(P_m W)$ as $m \to \infty$, and consequently, it is compact itself.

But this shows that R is a right Fredholm regularizer for A. In a symmetric manner one can check that $L := S^{-1}(I - WB)$ is a left Fredholm regularizer of A. $\qquad\square$

Figure 1 on page 15 and Figure 4 on page 57 also nicely illustrate another undisputable fact. In the perfect case $1 < p < \infty$ with dim $\mathbf{X} < \infty$, where much of the theory of this book grew up (for instance, see [67]), everything is rather nice:

$$
\Big(K(E, \mathcal{P}) \; ; \; L(E, \mathcal{P}) \; ; \; \text{invertibility at } \infty \Big) = \Big(K(E) \; ; \; L(E) \; ; \; \text{Fredholmness} \Big)
$$

In fact, no one of the left hand side items even appears in [67]. The bifurcation of the theory starts when we leave that case, and it culminates when $p = 1$ or $p = \infty$ and dim $\mathbf{X} = \infty$. It turns out that the appropriate way to follow at this bifurcation point is the left one, and that the above equality is just a coincidence in the case $1 < p < \infty$, dim $\mathbf{X} < \infty$.

2.4 Invertibility at Infinity vs. Stability

In Section 1.7 we have seen that one of the main ingredients of the applicability of an approximation method is its stability. The aim of this section is to translate the stability problem for a given approximation method into the question whether or not an associated operator is invertible at infinity.

Throughout the following, let $n \in \mathbb{N}$, $p \in [1, \infty]$, and let E denote either the space $L^p = L^p(\mathbb{R}^n)$ or $\ell^p(\mathbb{Z}^n, \mathbf{X})$ where $\mathbf{X}$ is an arbitrary Banach space. Depending on the choice of E; that is, whether we are in the continuous case or in the discrete case, take $\mathbb{D} \in \{\mathbb{R}, \mathbb{Z}\}$ corresponding, and define $E' = L^p(\mathbb{R}^{n+1})$ or $E' = \ell^p(\mathbb{Z}^{n+1}, \mathbf{X})$, respectively. To make a distinction between operators on E and operators on E', we will write I', P'_U, P'_k, Q'_U, Q'_k for the identity operator and the respective projection operators on E'.

2.4.1 Stacked Operators

Definition 2.16. *Given an $u \in E'$, we define the sequence $(u_\tau)_{\tau \in \mathbb{D}}$ with $u_\tau \in E$ by*

$$u_\tau(x) := u(x, \tau), \qquad x \in \mathbb{D}^n, \ \tau \in \mathbb{D},$$

and regard u_τ as the τ-th layer of u.

In this sense we will henceforth think of elements $u \in E'$ as being composed by their layers $u_\tau \in E$, $\tau \in \mathbb{D}$, where we will treat the whole sequence $(u_\tau)_{\tau \in \mathbb{D}}$ as one object in

$$\ell^p(\mathbb{D}, E) \cong \ell^p(\mathbb{Z}^{n+1}, \mathbf{X}) = E' \qquad \text{or} \qquad L^p(\mathbb{D}, E) \cong L^p(\mathbb{R}^{n+1}) = E'. \qquad (2.9)$$

Remark 2.17. Note that, for every $\tau \in \mathbb{D}$, the set $\{(x, \tau) : x \in \mathbb{D}^n\}$ has measure zero in $\mathbb{D}^{n+1}$ if $\mathbb{D} = \mathbb{R}$, but it has a positive measure if $\mathbb{D} = \mathbb{Z}$. This leads to a slight bifurcation of the following theory for the continuous and the discrete case.

In the continuous case, $u \in E' = L^p(\mathbb{R}^{n+1})$, it makes no sense to talk about a single layer $u_\tau \in E$ of u since u_τ is defined on a set of measure zero in $\mathbb{R}^{n+1}$, whereas in the discrete case, every single layer $u_\tau \in E$ of $u \in E'$ is well-defined since u_τ is defined on a set of positive measure in $\mathbb{Z}^{n+1}$. $\square$

We will use the layer construction from Definition 2.16 to associate an operator on E' with every bounded sequence $(A_\tau)_{\tau \in \mathbb{D}}$ of operators on E, simply by "stacking" this operator sequence to a "pile".

Definition 2.18. *If $T \subset \mathbb{D}$ and $(A_\tau)_{\tau \in T}$ is a bounded sequence of operators on E, then by*

$$(Bu)(x, \tau) := \begin{cases} (A_\tau u_\tau)(x) & \text{if } \tau \in T, \\ u_\tau(x) & \text{if } \tau \notin T, \end{cases} \qquad x \in \mathbb{D}^n, \ \tau \in \mathbb{D},$$

an operator B on E' is given, which acts as A_τ in the τ-th layer of u for $\tau \in T$, and it is the identity operator otherwise. In this sense, we will regard A_τ as the τ-th layer of B, and we will refer to B as the stacked operator of the sequence $(A_\tau)_{\tau \in T}$, denoted by

$$\bigoplus_{\tau \in T} A_\tau$$

or just by $\oplus A_\tau$ if the index set T is fixed.

By its definition, $\oplus A_\tau$ acts on every layer of $u \in E'$ independently from the other layers, like a generalized multiplication operator. If we identify u and (u_τ) in the sense of (2.9), then we can identify $\oplus A_\tau$ with the operator of multiplication by a function in

$$\ell^\infty(\mathbb{D}, L(E)) = \ell^\infty(\mathbb{Z}, L(E)) \qquad \text{or} \qquad L^\infty(\mathbb{D}, L(E)) = L^\infty(\mathbb{R}, L(E)), \qquad (2.10)$$

respectively. In the discrete case, this is a generalized multiplication operator in the sense of Definition 1.4. Identification (2.10) is the reason why, when we write about an 'essential supremum', about 'essential boundedness' or 'essential invertibility' in the following, for the discrete case this simply translates to the usual supremum, to 'uniform boundedness' and 'uniform invertibility', respectively.

By the identification (2.10), $\oplus A_\tau$ is a bounded linear operator on E' with

$$\left\| \bigoplus_{\tau \in T} A_\tau \right\|_{L(E')} = \begin{cases} s & \text{if } T = \mathbb{D}, \\ \max(s, 1) & \text{if } T \neq \mathbb{D}, \end{cases} \qquad \text{where} \qquad s = \operatorname*{ess\,sup}_{\tau \in T} \|A_\tau\|_{L(E)}.$$

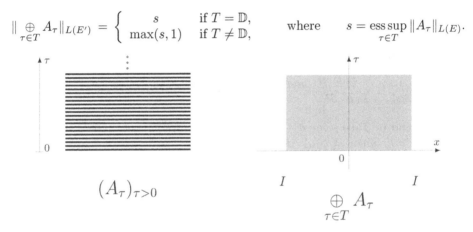

Figure 5: A rough illustration of the mapping $(A_\tau)_{\tau \in T} \mapsto \bigoplus_{\tau \in T} A_\tau$ for $n = 1$, $\mathbb{D} = \mathbb{R}$, $T = \mathbb{R}_+$.

Remark 2.19. By identifying $\oplus A_\tau$ with an operator of multiplication by a function in (2.10), it is clear that, in the continuous case, roughly speaking, the stacked operator $\oplus A_\tau$ doesn't even 'see' the behaviour of a single operator A_τ of the sequence. This is manifested by all the 'essential' notations in connection with $\oplus A_\tau$.   □

Proposition 2.20. *Let $(A_\tau)_{\tau \in T} \subset L(E)$ be a bounded sequence. The operator $\oplus A_\tau$ is invertible in $L(E')$ if and only if the set $\{A_\tau\}_{\tau \in T}$ is essentially invertible.*

Proof. This is an immediate consequence of Definition 1.57 and the identification of $\oplus A_\tau$ with an operator of multiplication by a function in (2.10).   □

In general, it is not true that $\oplus A_\tau \in BO$ or BDO^p or $L(E', \mathcal{P})$ whenever all operators A_τ are of that kind.

Example 2.21. a) Take $T = \mathbb{N}$ and $A_\tau = V_\tau$ for all $\tau \in T$. Then even $(A_\tau) \subset BO$, but it is readily seen that $\oplus A_\tau$ in not even in $L(E', \mathcal{P}) \supset BDO^p \supset BO$.

b) Take $T = \mathbb{R}_+$ and $A_\tau = V_{1/\tau}$ for all $\tau \in T$. Also in this case $(A_\tau) \subset \mathrm{BO}$ and even $A_\tau \xrightarrow{P} I$ as $\tau \to \infty$ if $p < \infty$. But still $\oplus A_\tau \notin L(E', \mathcal{P})$ holds.   □

But one can prove that $\oplus A_\tau$ is in BO or BDO^p or $L(E', \mathcal{P})$ if all A_τ are in the respective class, and, in addition, some uniformity condition is fulfilled. Recall that by $f_A(d)$ we denote the information flow (1.30) of A over the distance d.

Proposition 2.22. *Suppose* $(A_\tau)_{\tau \in T} \subset L(E)$ *is a bounded sequence.*

a) *If* $(A_\tau)_{\tau \in T} \subset \mathrm{BO}$ *and the band-widths of* A_τ *are essentially bounded, then* $\oplus A_\tau$ *is a band operator as well.*

b) *If* $(A_\tau)_{\tau \in T} \subset \mathrm{BDO}^p$ *and the condition*

$$\operatorname*{ess\,sup}_{\tau \in T} f_{A_\tau}(d) \to 0 \qquad as \qquad d \to \infty$$

holds, then $\oplus A_\tau$ *is band-dominated as well.*

c) *If* $(A_\tau)_{\tau \in T} \subset L(E, \mathcal{P})$ *and condition* (1.43) *holds; that is* $(A_\tau) \in \mathcal{F}(E, \mathcal{P})$, *then* $\oplus A_\tau \in L(E', \mathcal{P})$.

Proof. a) and b) Take some $d \in \mathbb{N}$ and two (measurable) sets $U, V \subset \mathbb{D}^{n+1}$ with $\operatorname{dist}(U, V) > d$. Then $P'_V (\oplus A_\tau) P'_U = \oplus C_\tau$, where

$$C_\tau = \begin{cases} P_{V^\tau} A_\tau P_{U^\tau} & \text{if } \tau \in T, \\ P_{V^\tau} P_{U^\tau} & \text{if } \tau \notin T, \end{cases}$$

with $U^\tau = \{x \in \mathbb{D}^n : (x, \tau) \in U\}$, $V^\tau = \{x \in \mathbb{D}^n : (x, \tau) \in V\}$ for $\tau \in \mathbb{D}$, and $P_\varnothing = 0$. But from

$$\operatorname{dist}(U_\tau, V_\tau) \ge \operatorname{dist}(U, V) > d, \qquad \tau \in \mathbb{D}$$

and $\|P'_V (\oplus A_\tau) P'_U\| = \operatorname{ess\,sup} \|C_\tau\|$, we get the claim in a) and b), respectively.

c) Take $k, m \in \mathbb{N}$, and note that $P'_k (\oplus A_\tau) Q'_m = \oplus C_\tau$ with

$$C_\tau = \begin{cases} P_k A_\tau Q_m & \text{if } \tau \in T \text{ and } |\tau| \le k, \\ 0 & \text{if } \tau \notin T \text{ or } |\tau| > k, \end{cases}$$

whenever $m > k$. Together with condition (1.43) we get that, for all $k \in \mathbb{N}$, $\|P'_k (\oplus A_\tau) Q'_m\| = \operatorname{ess\,sup} \|C_\tau\| \to 0$ as $m \to \infty$. Of course, the symmetric property, $\|Q'_m (\oplus A_\tau) P'_k\| \to 0$ as $m \to \infty$, is shown analogously.   □

Remark 2.23. Note that, for the usage in this proposition, the supremum in (1.43) can be reduced to an essential supremum.   □

2.4.2 Stability and Stacked Operators

Our intuition now tells us that the stability of the operator sequence is very closely related with the stacked operator's invertibility at infinity. But we shall see that, in the continuous case, these two properties are not the same.

For the following, let $T \subset \mathbb{D}$ be bounded below by 0, but unbounded towards plus infinity, whereby in the continuous case, we will in addition suppose that T is measurable and has a positive measure.

Proposition 2.24. *Consider a bounded sequence $(A_\tau)_{\tau \in T}$ in $L(E)$, almost every operator of which is invertible at infinity; that is*

$$A_\tau B_\tau = I + K_\tau \qquad and \qquad B_\tau A_\tau = I + L_\tau \tag{2.11}$$

for almost all $\tau \in T$, with some $B_\tau \in L(E)$ and $K_\tau, L_\tau \in K(E, \mathcal{P})$, and, in addition, (B_τ) is bounded, and, for every fixed $\tau_* \in T$,

$$\operatorname*{ess\,sup}_{\tau \leq \tau_*} \left\{ \|Q_m K_\tau\|, \|K_\tau Q_m\|, \|Q_m L_\tau\|, \|L_\tau Q_m\| \right\} \to 0 \qquad as \qquad m \to \infty. \tag{2.12}$$

Then the stacked operator $\oplus A_\tau$ is invertible at infinity if and only if there is some $\tau_* \in T$ such that the set $\{A_\tau\}_{\tau \in T, \tau > \tau_*}$ is essentially invertible.

Proof. Suppose there is a $\tau_* \in T$ such that the set $\{A_\tau\}_{\tau \in T, \tau > \tau_*}$ is essentially invertible. Then, for every $\tau \in T$, put

$$R_\tau := \begin{cases} A_\tau^{-1} & \text{if } A_\tau^{-1} \text{ exists,} \\ B_\tau & \text{otherwise.} \end{cases}$$

By the essential boundedness of the sets $\{A_\tau^{-1}\}_{\tau \in T, \tau > \tau_*}$ and $\{B_\tau\}_{\tau \in T}$, the operator $\oplus R_\tau$ is bounded. We show that $K := (\oplus A_\tau)(\oplus R_\tau) - I'$ is in $K(E', \mathcal{P})$. Clearly, $K = \oplus K_\tau$, where

$$K_\tau = \begin{cases} A_\tau R_\tau - I & \text{if } \tau \in T, \\ 0 & \text{if } \tau \notin T. \end{cases}$$

Take $\varepsilon > 0$ arbitrary, and choose $m \in \mathbb{N}$ such that $m > \tau_*$ and the essential supremum in (2.12) is less than ε. Then $Q_m' K = \oplus L_\tau$, where

$$L_\tau = \begin{cases} A_\tau R_\tau - I & \text{if } \tau \in T \text{ and } \tau > m, \\ Q_m(A_\tau R_\tau - I) & \text{if } \tau \in T \text{ and } \tau \leq m, \\ 0 & \text{if } \tau \notin T, \end{cases}$$

showing that, by the definition of R_τ, almost all layers L_τ with $\tau > \tau_*$ are zero, whence $\|Q_m' K\| = \operatorname{ess\,sup}_{\tau \leq \tau_*} \|Q_m(A_\tau R_\tau - I)\| < \varepsilon$. Analogously, we see that also $\|K Q_m'\| < \varepsilon$, showing that $K = (\oplus A_\tau)(\oplus R_\tau) - I' \in K(E', \mathcal{P})$. By the same argument one easily checks that also $(\oplus R_\tau)(\oplus A_\tau) - I' \in K(E', \mathcal{P})$, and hence that $\oplus A_\tau$ is invertible at infinity.

Now suppose that $\oplus A_\tau$ is invertible at infinity. Then, by Proposition 2.6, $\oplus A_\tau$ is weakly invertible at infinity, which means that there is an $m \in \mathbb{N}$ such that

$$Q'_m(\oplus A_\tau)B = Q'_m = C(\oplus A_\tau)Q'_m$$

holds with some $B, C \in L(E')$. If we put $U := \{(x, \tau) \in \mathbb{D}^{n+1} : x \in \mathbb{D}^n, \tau > m\}$ and multiply the left equality by P'_U from the left and the right equality from the right, we get

$$P'_U(\oplus A_\tau)B = P'_U = C(\oplus A_\tau)P'_U$$

since $P'_U Q'_m = P'_U = Q'_m P'_U$. But this equality clearly shows that $\oplus_{\tau > m} A_\tau$ is invertible, and hence the set $\{A_\tau\}_{\tau > m}$ is essentially invertible, by Proposition 2.20. $\square$

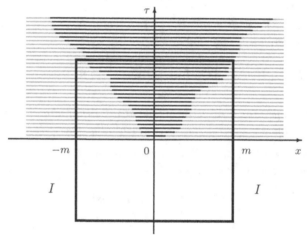

Figure 6: Illustration of the proof of Proposition 2.24 for the finite section method. If $\{A_\tau\}_{\tau > \tau_*}$ is essentially invertible, then everything outside a sufficiently large cube is (locally) invertible.

For example, if we consider the finite section method

$$A_\tau = A_{\lceil \tau \rceil} = P_{\Omega_\tau} A P_{\Omega_\tau} + Q_{\Omega_\tau}, \qquad \tau \in T \qquad (2.13)$$

for $A \in L(E)$ with a monotonously increasing sequence (Ω_τ) of bounded sets in $\mathbb{D}^n$, then the conditions of Proposition 2.24 are clearly fulfilled with $B_\tau = I$, and the essential supremum in (2.12) is zero as soon as

$$m \geq \sup_{x \in \Omega_{\tau_*}} |x|.$$

Corollary 2.25. *If* $(A_\tau) = (A_{\lceil \tau \rceil})$ *is the finite section sequence* (2.13) *for an operator* $A \in L(E)$, *then the stacked operator* $\oplus A_\tau$ *is invertible at infinity if and only if there is some* $\tau_* \in T$ *such that the set* $\{A_\tau\}_{\tau > \tau_*, \tau \in T}$ *is essentially invertible.*

What we have learnt from Proposition 2.24 and Corollary 2.25 is that, for very common classes of approximation methods (A_τ), the invertibility at infinity of the stacked operator $\oplus A_\tau$ is equivalent to the **essential** invertibility of the final part of the method. On the other hand, remember that the stability of a method is equivalent to the **uniform** invertibility of its final part.

Although this is tempting, there is absolutely no point in studying an approximation method, the final part of which is only essentially invertible. Let's call this property "essential stability" for a moment. The person, who encountered us with the equation $Au = b$ (1.33), will choose a subsequence $(A_{\tau_k})_{k=1}^\infty$ out of $(A_\tau)_{\tau \in T}$, hoping to approximate the solution u of (1.33) by the solutions u_k of $A_{\tau_k} u_k = b$ (1.35) as $k \to \infty$. However, if (A_τ) is only what we called "essentially stable", we can not guarantee that this works for **every** subsequence (A_{τ_k}) – only for **almost every one**. For instance, $(\tilde{A}_\tau)$ from Example 2.26 below is "essentially stable". But the final part of (τ_k) must not contain any integers to provide us with an applicable approximation method $(\tilde{A}_{\tau_k})$. So we better forget about "essential stability" and come back to our problem: stability and stacked operators.

Example 2.26. The sequence $(A_\tau)_{\tau \in \mathbb{R}_+}$ with $A_\tau = I$ for all $\tau \in \mathbb{R}_+$ is obviously stable, and $\oplus A_\tau = I'$ is clearly invertible at infinity. But perturbing this sequence in the integer components only,

$$\tilde{A}_\tau = \begin{cases} 0 & \text{if } \tau \in \mathbb{N}, \\ I & \text{otherwise,} \end{cases}$$

completely destroys the stability of the sequence, while it does not change the stacked operator's invertibility at infinity – not even the stacked operator itself which is still the identity I' on E'. □

The remainder of this chapter is devoted to the question under which conditions essential and uniform invertibility coincide, and hence, under which additional conditions Proposition 2.24 and Corollary 2.25 tell us that the sequence is stable if and only if the stacked operator is invertible at infinity.

The Discrete Case

In the discrete case $E = \ell^p(\mathbb{Z}^n, \mathbf{X})$, where we agreed to use a discrete index set $T \subset \mathbb{N}$, essential and uniform invertibility are clearly equivalent, as we remember from the discussion after (2.10). This fact allows us to state the following versions of Proposition 2.24 and Corollary 2.25 for the discrete case.

Corollary 2.27. *Suppose $E = \ell^p(\mathbb{Z}^n, \mathbf{X})$, $T \subset \mathbb{N}$, and $(A_\tau)_{\tau \in T}$ is a bounded sequence in $L(E)$, every operator of which is invertible at infinity; that is (2.11) holds with some $B_\tau \in L(E)$ and $K_\tau, L_\tau \in K(E, \mathcal{P})$, and, in addition, (B_τ) is bounded. Then the sequence (A_τ) is stable if and only if the stacked operator $\oplus A_\tau$ is invertible at infinity.*

Note that the set $\{\tau \in T : \tau \le \tau_*\}$ is finite for every $\tau_* \in T$, whence condition (2.12) automatically follows from $K_\tau, L_\tau \in K(E, \mathcal{P})$.

Theorem 2.28. *If $E = \ell^p(\mathbb{Z}^n, \mathbf{X})$, $T \subset \mathbb{N}$, and $(A_\tau) = (A_{\lceil \tau \rceil})$ is the finite section sequence (2.13) for an operator $A \in L(E)$, then the sequence (A_τ) is stable if and only if the stacked operator $\oplus A_\tau$ is invertible at infinity.*

The Continuous Case

Now suppose $E = L^p$ with $1 \le p \le \infty$. If we would restrict ourselves to sequences $(A_\tau)_{\tau \in T}$ with a discrete index set T, then Corollary 2.27 and Theorem 2.28 would immediately apply for this case as well. But we will now study the case of an index set $T \subset \mathbb{R}_+$ with positive measure in $\mathbb{R}$, which is often desirable for approximation methods in the continuous case, as we agreed before.

In Section 1.4 we have seen that, under several continuity assumptions on the mapping $\tau \mapsto A_\tau$, essential and uniform invertibility coincide. By Proposition 1.62 c) and d), we might want to expect the operators A_τ to depend **sufficiently smoothly** on τ. We will present three different approaches to this idea.

1st approach. In Lemma 1.61 we have shown that continuous dependence of A_τ on τ is sufficiently smooth. But this is one of the strongest forms of sufficient smoothness, and it is indeed not very fruitful to expect the mapping $\tau \mapsto A_\tau$ to be continuous from $\mathbb{R}_+$ to $L(E)$ since already very simple sequences like (P_τ) or the finite section sequence $(A_{\lceil \tau \rceil})$ for $A = 2I$ do not enjoy this property. The problem is that $\|P_\Omega - P_{\Omega'}\| = 1$ as soon as Ω and $\Omega' \subset \mathbb{R}^n$ differ in a set of positive measure.

A work-around for this problem, in connection with the finite section method, is as follows. Suppose the choice of the sets Ω_τ in (2.13) is such that there is a set $\Omega \subset \mathbb{R}^n$, and, for every $\tau \in T$, there is a bijective deformation φ_τ of $\mathbb{R}^n$ which sends Ω_τ to Ω. For example, if $\Omega_\tau = \tau\Omega$, then this deformation is just shrinking the $\mathbb{R}^n$ by the factor τ. Based on the deformations φ_τ, define invertible isometries Φ_τ on E. For example, if $\varphi_\tau(x) = x/\tau$ is the shrinking map mentioned above, then one can put $(\Phi_\tau u)(x) = \rho_\tau u(x/\tau)$, where

$$\rho_\tau = \begin{cases} \tau^{-n/p}, & p < \infty, \\ 1, & p = \infty \end{cases}$$

is a norm correction factor. Now transform all operators $A_\tau = A_{\lceil \tau \rceil}$ in (2.13) from operators which essentially act on $L^p(\Omega_\tau)$ to operators, say $A_\tau^\Omega := \Phi_\tau^{-1} A_\tau \Phi_\tau$, all essentially acting on the same space $L^p(\Omega)$, and then require continuous dependence of A_τ^Ω on τ.

With this construction we get that (A_τ) and (A_τ^Ω) are simultaneously stable, and $\oplus A_\tau$ and $\oplus A_\tau^\Omega$ are simultaneously invertible at infinity. So, under the assumption that $\tau \mapsto A_\tau^\Omega$ is continuous, we have that the sequence (A_τ) is stable if and only if $\oplus A_\tau$ is invertible at infinity.

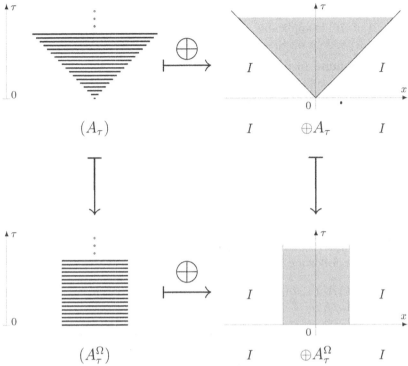

Figure 7: This figure illustrates the idea of the 1st approach. It shows the sequences (A_τ), (A_τ^Ω) and their stacked operators for $\Omega_\tau = \tau\Omega$ and $T = \mathbb{R}_+$.

The problem of this approach is that this continuity assumption on $\tau \mapsto A_\tau^\Omega$ is still a very strong restriction on the sequence (A_τ). For example, if $A \in L(E)$ is composed – via addition and composition – by a multiplication operator M_b and some other operators, and A_τ is the finite section sequence (2.13) of A, then this continuity condition requires the function $b \in L^\infty$ to be uniformly continuous (with some possible exception at the origin), which is rather restrictive!

2nd approach. Since the assumption that $A_\tau \rightrightarrows A_{\tau_0}$ as $\tau \to \tau_0$ turned out to be a bit restrictive, we consider a weaker type of continuity now. From Proposition 1.1.21 in [70] and the discussion afterwards, we conclude that, for $(A_\tau) \in \mathcal{F}(E, \mathcal{P})$, the mapping $\tau \mapsto A_\tau$ is sufficiently smooth if $A_\tau \xrightarrow{\mathcal{P}} A_{\tau_0}$ for $\tau \to \tau_0$. But it is readily seen that this is, in general, not the case since, again, the condition already fails for $(A_\tau) = (P_{\Omega_\tau})$, and hence for the finite section method.

Again, there is a work-around. For every $\tau_0 \in T$, one can define an approximate identity $\tilde{\mathcal{P}} = (\tilde{P}_m)_{m=1}^\infty$ such that $P_{\Omega_\tau} \xrightarrow{\tilde{\mathcal{P}}} P_{\Omega_{\tau_0}}$ as $\tau \to \tau_0$. For example, if Ω is compact and convex and $\Omega_\tau = \tau\Omega$, one can choose

$$\tilde{P}_m = I - P_{(1+1/m)\tau_0\Omega} + P_{(1-1/m)\tau_0\Omega}, \qquad m \in \mathbb{N}.$$

However, due to this new approximate identity, the class $\mathcal{F}(E, \tilde{P})$, for which Proposition 1.1.21 of [70] is applicable, completely differs from $\mathcal{F}(E, \mathcal{P})$, and, moreover, this class is dependent on τ_0.

3rd approach. Instead of $A_\tau \rightrightarrows A_{\tau_0}$ or $A_\tau \xrightarrow{\mathcal{P}} A_{\tau_0}$, we will now, depending on $E = L^p$, either suppose $*$-strong convergence $A_\tau \xrightarrow{*} A_{\tau_0}$ or pre$*$-strong convergence $A_\tau \xrightarrow{\lhd} \cdot A_{\tau_0}$ as $\tau \to \tau_0$. Precisely, for every $\tau_0 \in T$, we expect the strong convergence

$$A_\tau \to A_{\tau_0} \tag{2.14}$$

on E and the strong convergence of the (pre-)adjoints

$$A_\tau^* \to A_{\tau_0}^* \quad \text{if} \quad p < \infty \qquad\qquad \text{or} \qquad\qquad A_\tau^\lhd \to A_{\tau_0}^\lhd \quad \text{if} \quad p = \infty \tag{2.15}$$

on the (pre-)dual space of E as $\tau \to \tau_0$.

This type of continuous dependence of A_τ on τ is sufficiently smooth, as we will prove in the following lemma.

Lemma 2.29. *If $(A_\tau) \subset L(E)$ is subject to conditions (2.14) and (2.15), then the mapping $\tau \mapsto A_\tau$, from $T \subset \mathbb{R}$ to $L(E)$, is sufficiently smooth.*

Proof. Take an arbitrary $\tau_0 \in T$ and some sequence $\tau_k \to \tau_0$ in T as $k \to \infty$, and suppose the sequence $(A_{\tau_k})_{k=1}^\infty$ is stable. What we have to show is that then A_{τ_0} is invertible with $\|A_{\tau_0}^{-1}\| \le s := \sup_{k > k_*} \|A_{\tau_k}^{-1}\|$.

If $p < \infty$, then, by (2.14) and (2.15), we have $A_{\tau_k} \xrightarrow{*} A_{\tau_0}$ as $k \to \infty$. We will (independently from this lemma) show in Corollary 2.37 that, under these conditions, A_{τ_0} is invertible, and $\|A_{\tau_0}^{-1}\| \le s$.

If $p = \infty$, then, with (A_{τ_k}), also the sequence $(A_{\tau_k}^\lhd)$ is stable. By (2.14) and (2.15), we have $A_{\tau_k}^\lhd \xrightarrow{*} A_{\tau_0}^\lhd$ as $k \to \infty$. Now apply the same Corollary 2.37 to $(A_{\tau_k}^\lhd)$, showing that $A_{\tau_0}^\lhd$ is invertible, and $\|(A_{\tau_0}^\lhd)^{-1}\| \le s$. But then, also A_{τ_0} is invertible, and $\|A_{\tau_0}^{-1}\| \le s$. $\qquad\square$

Under some conditions on the sets Ω_τ, the finite section method (2.13) has both properties (2.14) and (2.15), and hence it is sufficiently smooth, showing that it is stable if and only if its stacked operator is invertible at infinity. We will henceforth study such a situation.

Let Ω be a compact and convex subset of $\mathbb{R}^n$, containing the origin in its interior. In what follows, we will choose the index set $T := \mathbb{R}_+$, and the sequence of sets Ω_τ from Section 1.5.3 shall be in the following convenient way:

$$\Omega_\tau := \tau\Omega, \qquad \tau > 0.$$

Then $\Omega_{\tau_1} \subset \Omega_{\tau_2}$ if $0 < \tau_1 < \tau_2$, and for every $x \in \mathbb{R}^n$, there exists a $\tau > 0$ such that $x \in \Omega_\tau$.

The case $1 < p < \infty$. Consider the finite section method (2.13) on $E = L^p$, where $1 < p < \infty$. It is easily seen that this method is subject to (2.14) and (2.15).

Proposition 2.30. *Let $E = L^p$ with $1 < p < \infty$. For every $A \in L(E)$, the finite sections $A_{\lceil \tau \rceil}$ depend sufficiently smoothly on $\tau \in \mathbb{R}_+$.*

Proof. For $1 < p < \infty$, the sequence of finite section projectors $P_{\tau\Omega}$ is subject to (2.14) and (2.15), where $P^*_{\tau\Omega}$ is equal to $P_{\tau\Omega}$, acting on $L^q \cong E^*$ with $1/p + 1/q = 1$. Since $*$-strong convergence is compatible with addition and multiplication, we get that also $(A_\tau) = (A_{\lceil \tau \rceil})$ is subject to (2.14) and (2.15), and it remains to apply Lemma 2.29. □

The cases L^1 and L^∞. Now we study the finite section method (2.13) on $E = L^p$ for $p = 1$ and for $p = \infty$. Our aim is to prove some analogue of Proposition 2.30. But clearly, for $p = 1$, we only have (2.14), and for $p = \infty$, we only have (2.15). So, recalling the proof of Lemma 2.29, if $\tau_k \to \tau_0 > 0$ and $(A_{\lceil \tau_k \rceil})$ is stable, then in either case, we can, roughly speaking, only prove 50% of the invertibility of $A_{\lceil \tau_0 \rceil}$.

Fortunately, the operators A that we have in mind in Section 4.2, have the property that a finite section $A_{\lceil \tau_0 \rceil}$ is either invertible or not even "semi-invertible". This is the theory we are going to develop now. Firstly, here is what we mean by "semi-invertible".

Definition 2.31. *We say that an operator $A \in L(E)$ is bounded below if the quotient $\|Ax\|/\|x\|$ is bounded below by a positive constant for all $x \in E \setminus \{0\}$. In that case, the best possible lower bound*

$$\nu(A) \;:=\; \inf_{x \in E \setminus \{0\}} \frac{\|Ax\|}{\|x\|} \;=\; \inf_{x \in E, \, \|x\|=1} \|Ax\| \;>\; 0$$

is referred to as the lower norm of A.

Lemma 2.32. *An operator $A \in L(E)$ is bounded below if and only if A is injective and has a closed image.*

Proof. If A is bounded below, then, clearly, $\ker A = \{0\}$, and every Cauchy sequence in $\operatorname{im} A$ converges in $\operatorname{im} A$.

Conversely, if A is injective and $\operatorname{im} A$ is closed, then Banach's theorem on the inverse operator says that the operator $Ax \mapsto x$ from $\operatorname{im} A$ to E is bounded. □

In fact, being bounded below is closely related with being one-sided invertible.

Lemma 2.33. *An operator $A \in L(E)$ is left invertible; that is $BA = I$ for some $B \in L(E)$, if and only if A is bounded below and $\operatorname{im} A$ is complementable in E.*

Proof. Suppose A is bounded below and $\operatorname{im} A$ is complementable. From Lemma 2.32 we know that A is injective and $\operatorname{im} A$ is closed. Let $C : \operatorname{im} A \to E$ act by $Ax \mapsto x$. By Banach's theorem on the inverse operator, C is bounded. Let B act like C on im A and as the zero operator on its complement. Then $B \in L(E)$ and $BA = I$.

If there is a $B \in L(E)$ with $BA = I$, then from $\|x\| = \|BAx\| \leq \|B\|\|Ax\|$ we see that $\|Ax\|/\|x\|$ is bounded below by $1/\|B\|$. Moreover, it turns out that ker B is a direct complement of im A in E. Indeed, every $x \in E$ can be written as $x = ABx + (x - ABx)$ where $ABx \in$ im A and $x - ABx \in$ ker B since $B(x - ABx) = Bx - (BA)Bx = 0$. Finally, suppose $y \in$ im $A \cap$ ker B. Then $y = Ax$ for some $x \in E$ and $By = 0$. Consequently, $0 = By = BAx = x$ which shows that $y = Ax = 0$. $\square$

Remark 2.34. If E is a Hilbert space, then im A is automatically complementable in E if A is bounded below. This is the case since every closed subspace of a Hilbert space has a complement (its orthogonal complement, for example). Consequently, if E is a Hilbert space, then $A \in L(E)$ is left invertible if and only if it is bounded below. $\square$

Lemma 2.35. $A \in L(E)$ *is invertible if and only if* A *and* A^* *are bounded below. In that case,* $\nu(A) = 1/\|A^{-1}\|$.

Proof. If $A \in L(E)$ is invertible, then also $A^* \in L(E^*)$ is invertible, and, by Lemma 2.33, both are bounded below. Conversely, if A and A^* are bounded below, then, by Lemma 2.32, A is injective and has a closed image. Since, by Lemma 2.32 again, also A^* is injective, the image of A is dense in E, and hence it is all of E. Consequently, A is invertible.

If $x \in E$, then $\|x\| = \|A^{-1}Ax\| \leq \|A^{-1}\|\|Ax\|$ shows $\|Ax\|/\|x\| \geq 1/\|A^{-1}\|$ if $x \neq 0$. This inequality can be made as tight as desired by the choice of $x \in E$. $\square$

Proposition 2.36. $(A_m)_{m=1}^{\infty} \subset L(E)$ *be a stable sequence with strong limit* $A \in L(E)$. *Then* A *is bounded below, and* $\nu(A) \geq 1/\sup_{m \geq m_*} \|A_m^{-1}\|$.

Proof. Take $m_* \in \mathbb{N}$ large enough that A_m is invertible for all $m \geq m_*$, and put $s := \sup_{m \geq m_*} \|A_m^{-1}\| < \infty$. Then, for every $x \in E$ and $m \geq m_0$, we have

$$\|x\| = \|A_m^{-1} A_m x\| \leq \|A_m^{-1}\| \|A_m x\| \leq s\|A_m x\| \to s\|Ax\| \qquad \text{as} \qquad m \to \infty.$$

So $\|Ax\|/\|x\|$ is bounded below by $1/s$. $\square$

Corollary 2.37. *If* $(A_m) \subset L(E)$ *is stable and* $A_m \xrightarrow{*} A$, *then* A *is invertible, and* $\|A^{-1}\| \leq \sup_{m \geq m_*} \|A_m^{-1}\|$.

Proof. If $(A_m) \subset L(E)$ is stable, then also $(A_m^*) \subset L(E^*)$ is stable. Now apply Proposition 2.36 and Lemma 2.35. $\square$

Lemma 2.38. *The map* $A \mapsto \nu(A)$ *is continuous in* $L(E) \to \mathbb{R}$; *precisely, for all* $A, B \in L(E)$, *it holds that* $|\nu(A) - \nu(B)| \leq \|A - B\|$.

Proof. Let $A, B \in L(E)$. Then, for all $x \in E$ with $\|x\| = 1$,

$$\|A - B\| \geq \|Ax - Bx\| \geq \|Ax\| - \|Bx\| \geq \nu(A) - \|Bx\|$$

holds. But the right hand side comes arbitrarily close to $\nu(A) - \nu(B)$ by choosing x accordingly. The other inequality, $\|A - B\| \geq \nu(B) - \nu(A)$, holds as well, by reasons of symmetry. □

Corollary 2.39. *The set of operators $A \in L(E)$ that are bounded below is an open subset of $L(E)$; precisely, if $A \in L(E)$ is bounded below, then every operator $B \in L(E)$ with $d := \|A - B\| < \nu(A)$ is bounded below, and $\nu(B) \geq \nu(A) - d > 0$.*

Definition 2.40. *Let $\Phi_+(E)$ denote the set of all operators $A \in L(E)$ which have a finite-dimensional kernel and a closed image. Operators in $\Phi_+(E)$ are called semi-Fredholm operators.*

Note that, by Lemma 2.32, if $A \in L(E)$ is bounded below, then $A \in \Phi_+(E)$. In analogy to Fredholmness, also semi-Fredholmness has the following properties.

Lemma 2.41. *Semi-Fredholmness is invariant*

a) *under small perturbations, and*

b) *under compact perturbations.*

Proof. See [9], for example. □

For the remainder of this section, let us decompose the set $L(E)$ into the three distinct subsets $\mathbf{i}(E)$, $\mathbf{b}(E)$ and $\mathbf{n}(E)$, namely the sets of operators that are

- invertible,

- bounded below but not invertible, and

- not bounded below,

respectively. Following our objective to study operators, the finite sections of which are either **invertible** or **not** even "semi-invertible", we will define a new set of operators by taking away from $L(E)$ all the operators that are bounded below but not invertible. So let

$$\mathbf{r}(E) \quad := \quad L(E) \setminus \mathbf{b}(E) \quad = \quad \mathbf{i}(E) \cup \mathbf{n}(E).$$

Then $\mathbf{r}(E)$ is the set of all operators $A \in L(E)$ with the property

$$A \text{ is invertible} \quad \textbf{or} \quad A \text{ is not even bounded below,}$$

or, in an equivalent formulation,

$$\textbf{If} \quad A \text{ is bounded below,} \quad \textbf{then} \quad A \text{ is even invertible.} \qquad (2.16)$$

For example, it is readily seen that all multiplication operators are subject to (2.16):

Lemma 2.42. *Take an arbitrary $b \in L^\infty$. If $M_b \in \Phi_+(E)$, then M_b is invertible.*

Proof. If $M_b \in \Phi_+(E)$, then, by Lemma 2.41 a), there is an $\varepsilon_0 > 0$ such that all operators A with $\|M_b - A\| < \varepsilon_0$ are in $\Phi_+(E)$ as well. Suppose M_b were not invertible. Then, for every $\varepsilon > 0$, there exists a set $S_\varepsilon \subset \mathbb{R}^n$ with positive measure such that $|b(x)| < \varepsilon$ for all $x \in S_\varepsilon$. Now put

$$c(x) := \left\{ \begin{array}{ll} b(x), & x \notin S_{\varepsilon_0/2}, \\ 0, & x \in S_{\varepsilon_0/2}. \end{array} \right.$$

Then $\|b - c\|_\infty < \varepsilon_0$, whence $M_c \in \Phi_+(E)$. But this contradicts the fact that M_c has an infinite dimensional kernel (c vanishes on all of $S_{\varepsilon_0/2}$). $\square$

Corollary 2.43. *The subalgebra $\{M_b : b \in L^\infty\}$ of all multiplication operators in $L(E)$ is contained in $\mathbf{r}(E)$.*

Proof. Take an arbitrary function $b \in L^\infty$. We show that M_b is subject to (2.16). Suppose M_b is bounded below. Then, by Lemma 2.32, $M_b \in \Phi_+(E)$. But now Lemma 2.42 shows that M_b is invertible. $\square$

Proposition 2.44. *The subset $\mathbf{r}(E)$ of $L(E)$ is closed under norm convergence and multiplication.*

Proof. **Norm convergence:** Take a sequence $(A_m)_{m=1}^\infty \subset \mathbf{r}(E)$ with $A_m \rightrightarrows A$ as $m \to \infty$. We will show that also $A \in \mathbf{r}(E)$, i.e. A has property (2.16).

If A is bounded below, then $M := \nu(A) > 0$. We choose some $m \in \mathbb{N}$ such that $d := \|A - A_m\| < M/3$. From Corollary 2.39 we know that then A_m is bounded below with

$$\|A_m x\| \geq (M - d)\|x\| > (M - \frac{M}{3})\|x\| = \frac{2}{3}M\|x\|, \qquad x \in E.$$

Since $A_m \in \mathbf{r}(E)$, we can, by (2.16), conclude that A_m is even invertible, and Lemma 2.35 tells that

$$\|A_m^{-1}\| \leq \frac{3}{2M}.$$

Consequently, we know that

$$\|A_m^{-1}(A - A_m)\| \leq \|A_m^{-1}\| \, \|A - A_m\| \leq \frac{3}{2M} \cdot \frac{M}{3} = \frac{1}{2},$$

and hence,

$$A = A_m \left(I + A_m^{-1}(A - A_m) \right)$$

is invertible. So the norm limit A has property (2.16) as well.

Multiplication: It is readily seen that, for all $A_i, B_i \in \mathbf{i}(E)$ and $A_n, B_n \in \mathbf{n}(E)$, $A_i B_i \in \mathbf{i}(E)$, $A_i B_n \in \mathbf{n}(E)$, $A_n B_i \in \mathbf{n}(E)$ and $A_n B_n \in \mathbf{n}(E)$ hold, and hence, $AB \in \mathbf{r}(E) = \mathbf{i}(E) \cup \mathbf{n}(E)$ if $A, B \in \mathbf{r}(E) = \mathbf{i}(E) \cup \mathbf{n}(E)$. $\square$

Remark 2.45. Note that $\mathbf{r}(E)$ is not closed under addition. For example, take $n = 1$, $p \in [1, \infty]$, and

$$A = V_1 P_{[0,+\infty)} \qquad \text{and} \qquad B = P_{(-\infty,0)}.$$

Then $A, B \in \mathbf{n}(E)$ but $A + B \in \mathbf{b}(E)$. $\qquad\qquad\square$

In accordance with our objective to study operators whose finite sections have property (2.16), we define the following two classes, where we abbreviate $(A_{\lceil \tau \rfloor})^{\triangleleft} = (A^{\triangleleft})_{\lceil \tau \rfloor}$ by $A^{\triangleleft}_{\lceil \tau \rfloor}$.

$$\mathcal{R}_1 \quad := \quad \left\{ A \in L(L^1) \ : \ A_{\lceil \tau \rfloor} \in \mathbf{r}(L^1) \quad \forall \tau > 0 \right\} \subset L(L^1),$$

$$\mathcal{R}_\infty \quad := \quad \left\{ A \in \mathcal{S} \ : \ A^{\triangleleft}_{\lceil \tau \rfloor} \in \mathbf{r}(L^1) \quad \forall \tau > 0 \right\} \subset L(L^\infty).$$

Now finally, we have found the right framework for our analogous formulation of Proposition 2.30 concerning the finite section method in L^1 and L^∞.

Proposition 2.46. *Let $p \in \{1, \infty\}$, $E = L^p$, and A be an arbitrary operator in $\mathcal{R}_p$. Then the finite sections $A_{\lceil \tau \rfloor}$ depend sufficiently smoothly on $\tau \in \mathbb{R}_+$.*

Proof. Take an arbitrary $\tau_0 > 0$ and a sequence $\tau_k \to \tau_0$ as $k \to \infty$, and suppose $(A_{\lceil \tau_k \rfloor})$ is stable. We have to show that $A_{\lceil \tau_0 \rfloor}$ is invertible in $L(E)$ with $\|A_{\lceil \tau_0 \rfloor}^{-1}\| \leq s := \sup \|A_{\lceil \tau \rfloor}^{-1}\|$.

Start with $p = 1$. Here we have strong convergence $P_{\tau_k \Omega} \to P_{\tau_0 \Omega}$, and hence

$$A_{\lceil \tau_k \rfloor} = P_{\tau_k \Omega} A P_{\tau_k \Omega} + Q_{\tau_k \Omega} \ \to \ P_{\tau_0 \Omega} A P_{\tau_0 \Omega} + Q_{\tau_0 \Omega} = A_{\lceil \tau_0 \rfloor}$$

strongly as $k \to \infty$. Applying Proposition 2.36 to the stable sequence $(A_{\lceil \tau_k \rfloor})$, we get that $A_{\lceil \tau_0 \rfloor}$ is bounded below and $\nu(A_{\lceil \tau_0 \rfloor}) \geq 1/s$. The condition $A \in \mathcal{R}_1$ now yields that $A_{\lceil \tau_0 \rfloor} \in \mathbf{r}(E)$, and hence, we can conclude from $A_{\lceil \tau_0 \rfloor}$ being bounded below to even being invertible. The inequality $\|A_{\lceil \tau_0 \rfloor}^{-1}\| = 1/\nu(A_{\lceil \tau_0 \rfloor}) \leq s$ follows.

For $p = \infty$, we pass to the sequence of pre-adjoints $(A^{\triangleleft}_{\lceil \tau_k \rfloor})$, which is stable as well, and apply the result for $p = 1$, yielding that $A^{\triangleleft}_{\lceil \tau_0 \rfloor}$, and hence $A_{\lceil \tau_0 \rfloor}$, is invertible and the desired norm inequality holds. $\qquad\square$

So – under somewhat stronger assumptions on A – in L^1 and L^∞, we have the same result as Proposition 2.30 in L^p with $1 < p < \infty$. In Section 4.2.2 we will see that $\mathcal{R}_1$ and $\mathcal{R}_\infty$ still contain sufficiently many interesting and practically relevant operators.

Now here is the outcome of our investigations on the finite section method in the continuous case which, together with Theorem 2.28 for the discrete case, completes our studies on the finite section method for this chapter. We will come back to the finite section method in Chapter 4.

Theorem 2.47. *If A is an operator in $L(L^p)$ with $1 < p < \infty$ or in $\mathcal{R}_1$ or $\mathcal{R}_\infty$, then the finite section method $(A_{\lceil \tau \rfloor})$ is stable if and only if $\oplus A_{\lceil \tau \rfloor}$ is invertible at infinity.*

Proof. By Corollary 2.25, the stacked operator $\oplus A_{\lceil\tau\rfloor}$ is invertible at infinity if and only if there is some $\tau_* > 0$ such that $\{A_{\lceil\tau\rfloor}\}_{\tau > \tau_*}$ is essentially invertible.

Propositions 2.30 and 2.46 show that, under the conditions of this theorem, the mapping $\tau \mapsto A_{\lceil\tau\rfloor}$ is sufficiently smooth. Moreover, $T = \mathbb{R}_+$ is massive, and Proposition 1.62 c) and d) shows that in this case the essential invertibility of $\{A_{\lceil\tau\rfloor}\}_{\tau > \tau_*}$ is equivalent to its uniform invertibility which is nothing but the stability of $(A_{\lceil\tau\rfloor})$. $\square$

2.5 Comments and References

The theory of Fredholm operators can be found in a huge number of books on operator theory, for example [31]. This theory essentially generalizes E. I. FRED-HOLM's classical results for operators of the form $I + K$ with $K \in K(E)$. In this connection, the name "Fredholm operator" originated. Sometimes these operators are also referred to as Noether or Φ-operators. F. NOETHER was the first who studied the more general situation of operators with a closed image and finite-dimensional kernel and cokernel [56]. The usage of the letter Φ, which is standard today, started with a mistranslation of the cyrillic letter "Ф" (for "Fredholm") from a publication of GOHBERG and KREIN.

The study of $\mathcal{P}$-Fredholmness and invertibility at infinity goes back to [68] and [70]. Proposition 2.2 is basically from [70]. The main part of Proposition 2.10 goes back to [68]. Proposition 2.12 is the product of personal communication with BERND SILBERMANN. Proposition 2.15 generalizes an argument from the proof of Proposition 3.3.4 in [70]. An even stronger version of Proposition 2.15 can be found in Section 3.3 of [17].

The idea to identify the stability of a sequence of operators with properties of an associated block-diagonal operator can already be found at GORODETSKI [32]. The results presented here for the discrete case are well-known (see [68] and [70], for example). The treatment of methods with a continuous index set, as presented here, goes back to [68] (the "1st approach") and [48], [50] (the "3rd approach"). Operators that are bounded below are well-studied in operator theory.

Chapter 3

Limit Operators

In Chapter 2 we introduced and discussed the property of invertibility at infinity. In particular, we made very detailed studies of its intimate connections with Fredholmness in Section 2.3, and we explored the possibility to study approximation methods and their stability as a special invertibility problem at infinity in Section 2.4, which opens the door to a variety of applications. So invertibility at infinity turns out to be the key property to be studied in this book. It is the aim of this chapter to introduce a tool for the study of this key property: *limit operators*.

Let $E = \ell^p(\mathbb{Z}^n, \mathbf{X})$ with $p \in [1, \infty]$ and a Banach space $\mathbf{X}$, and take an arbitrary operator $A \in L(E)$. In Example 2.7 a)–d) we have seen operators $A \in L(E)$ which are not surjective or not injective at infinity, respectively. Is there a generic way to detect such behaviour?

A First Glimpse at Localization

If A is a band-dominated operator, then Proposition 2.10 relates the question whether or not A is invertible at infinity to an invertibility problem in the Banach algebra $\mathrm{BDO}^p/K(E, \mathcal{P})$. For the study of invertibility in a Banach algebra $\mathbf{B}$, it is often a good idea to look at so-called *unital homomorphisms*; that is a mapping φ from $\mathbf{B}$ into another Banach algebra $\mathbf{A}$ which is subject to

$$\varphi(a \dotplus b) = \varphi(a) \dotplus \varphi(b) \quad \text{and} \quad \varphi(e_{\mathbf{B}}) = e_{\mathbf{A}}$$

for all $a, b \in \mathbf{B}$. Clearly, if an element $a \in \mathbf{B}$ is invertible in $\mathbf{B}$, then also $\varphi(a)$ is necessarily invertible in $\mathbf{A}$, and $\varphi(a)^{-1} = \varphi(a^{-1})$. In short,

$$a \text{ is invertible} \quad \Longrightarrow \quad \varphi(a) \text{ is invertible}, \quad \forall \text{ unital homomorphisms } \varphi. \quad (3.1)$$

The following example is about the mother of all commutative Banach algebras.

Example 3.1. Let $\mathbf{B} = C[0, 1]$ be the Banach algebra of all continuous functions on the interval $[0, 1]$, equipped with pointwise operations and the maximum norm. Moreover, let $\mathbf{A} = \mathbb{C}$, and fix a point $x \in [0, 1]$. We look at the mapping φ_x that assigns to every continuous function a on $[0, 1]$ its function value $a(x)$ at x.

So we have

$$\varphi_x : \mathbf{B} \to \mathbf{A}, \qquad \varphi_x : a \mapsto a(x).$$

Clearly, φ_x is a unital homomorphism. It is compatible with addition and multiplication, and it maps the function that is equal to 1 on all of $[0, 1]$ to $1 \in \mathbb{C}$. And indeed, if a is invertible as a function in $\mathbf{B}$; that is $a \in \mathbf{B}$ and $1/a \in \mathbf{B}$, then its function value at x is necessarily non-zero; that is $\varphi_x(a)$ is invertible in $\mathbf{A} = \mathbb{C}$. $\qquad\qquad\qquad\qquad\qquad\qquad\qquad\qquad\qquad\qquad\qquad\qquad\quad\square$

If we no longer fix the point $x \in [0, 1]$ in Example 3.1, then (3.1) tells us that

$$a \text{ is invertible} \quad \Longrightarrow \quad \varphi_x(a) \text{ is invertible}, \qquad \forall x \in [0, 1]. \qquad (3.2)$$

Our knowledge about continuous functions on compact sets tells us that the invertibility of all elements $\varphi_x(a)$ is not only necessary for the invertibility of a in $\mathbf{B}$, as (3.1) and (3.2) say, but also sufficient! That is what we call a *sufficient family* of unital homomorphisms. Our family of unital homomorphisms $\{\varphi_x\}_{x \in [0,1]}$ contains enough elements to even yield the other implication in (3.2), because, roughly speaking, it simply looks everywhere.

Despite its simplicity, Example 3.1 did not only lead us to the notion of a sufficient family, it also nicely illustrates the main idea of local principles in Banach algebras:

Replace a difficult invertibility problem in $\mathbf{B}$
by a family of simpler invertibility problems.

Very often this is done by, again speaking roughly, looking at $a \in \mathbf{B}$ through a family of microscopes. In Example 3.1 we might think of a little microscope attached to every point $x \in [0, 1]$, and the process of looking at the separate elements $\{\varphi_x(a)\}_{x \in [0,1]} \subset \mathbf{A}$ rather than at the whole element $a \in \mathbf{B}$ at once, is referred to as *localization* over $[0, 1]$.

If we want to investigate the invertibility problem that is connected with invertibility at infinity of an operator A, there is clearly one place we have to look at: infinity[1].

The somewhat simpler objects, that we find there via application of an appropriate family of unital homomorphisms, each represent the local behaviour of A in a certain location at infinity. We could call them "infinite local representatives" or "samples at infinity" of A, but in a minute we will start calling them the limit operators of A.

---

[1] In that case, we might prefer telescopes instead of microscopes.

How to Pick Samples from Infinity – A Rough Guide

Given a point $\vartheta \in \mathbb{Z}^n$, one might ask how the action of our operator A on $u \in E$ looks like from another point of view, say ϑ taking the role of the origin. Well, for finite points ϑ, the answer is clearly $V_{-\vartheta} A V_{\vartheta}$ which is still a very close relative of the operator A itself. But what if we ask for the point $\vartheta = \infty$? Then ϑ can only be approached as the limit of a sequence h of finite points $h_1, h_2, \ldots$ in $\mathbb{Z}^n$, and by doing the above process for every one of these finite points h_m, the answer for $\vartheta = \infty$ can only be understood in the sense of some sort of limit of the sequence of operators $V_{-h_m} A V_{h_m}$ as m goes to infinity. This limit is what we will call a limit operator.

Obviously, there are many different paths $h = (h_1, h_2, \ldots)$ leading to infinity which, in general, produce many different limit operators of A. The behavior of A at $\vartheta = \infty$ will be reflected by the collection of all these limit operators of A.

3.1 Definition and Basic Properties

Let $E = \ell^p(\mathbb{Z}^n, \mathbf{X})$ with $p \in [1, \infty]$ and a Banach space $\mathbf{X}$. In order to derive a family of local representatives of $A \in L(E)$ at infinity, we need something like a partition of infinity into different locations. For $n = 1$, a rough partition of ∞ that we are very familiar with is $\{-\infty, +\infty\}$. We will generalize this idea of partitioning ∞ (with respect to the different directions in which it can be approached) to $\mathbb{R}^n$.

Definition 3.2. *Let S^{n-1} denote the unit sphere in $\mathbb{R}^n$ with respect to the Euclidian norm $|.|_2$. If $R > 0$, $s \in S^{n-1}$ and $V \subset S^{n-1}$ is a neighbourhood of s in S^{n-1}, then*

$$U_R^\infty := \{x \in \mathbb{R}^n \ : \ |x|_2 > R\} \tag{3.3}$$

is called a neighbourhood of ∞, and

$$U_{R,V}^\infty := \{x \in \mathbb{R}^n \ : \ |x|_2 > R \ \text{ and } \ \frac{x}{|x|_2} \in V\} \tag{3.4}$$

is a neighbourhood of ∞_s. We say that a sequence $(x_m)_{m=1}^\infty \subset \mathbb{R}^n$ tends to infinity or tends to infinity in direction s, and we write $x_m \to \infty$ or $x_m \to \infty_s$, if, for every neighbourhood U of ∞ or ∞_s, respectively, all but finitely many elements of the sequence (x_m) are in U.

If $n = 1$, we will, of course, use the familiar notations $-\infty$ and $+\infty$ instead of ∞_{-1} and ∞_{+1}. Since the Euclidian norm $|.|_2$ and the maximum norm $|.|$ on $\mathbb{R}^n$ are equivalent, we have that x_m tends to infinity if and only if $|x_m| \to \infty$ as $m \to \infty$.

Now finally, here is the definition of a limit operator.

Definition 3.3. *Let $A \in L(E)$, and $h = (h_m)$ be a sequence in $\mathbb{Z}^n$ tending to infinity. If the sequence of operators*

$$V_{-h_m} A V_{h_m}, \qquad m = 1, 2, \ldots \tag{3.5}$$

$\mathcal{P}$-converges to an operator $B \in L(E)$ as $m \to \infty$, then we call B the limit operator of A with respect to the sequence h, and denote it by A_h. In this case, we will say that the sequence h leads to a limit operator of A.

From the uniqueness of the $\mathcal{P}$-limit we know that a limit operator A_h is unique if it exists. If $h \subset \mathbb{Z}^n$ is a sequence tending to infinity such that A_h exists, and if g is an infinite subsequence of h, then also g leads to a limit operator of A, and $A_g = A_h$.

The passage to a limit operator is compatible with addition, composition, passing to the norm limit and to the adjoint operator.

Proposition 3.4. *Let A, B, $A^{(1)}$, $A^{(2)}, \ldots$ be arbitrary operators in $L(E, \mathcal{P})$, and let $h \subset \mathbb{Z}^n$ be a sequence tending to infinity.*

a) *If A_h exists, then $\|A_h\| \leq \|A\|$.*

b) *If A_h and B_h exist, then $(A + B)_h$ exists and is equal to $A_h + B_h$.*

c) *If A_h and B_h exist, then $(AB)_h$ exists and is equal to $A_h B_h$.*

d) *If $A^{(m)} \rightrightarrows A$ as $m \to \infty$, and the limit operators $(A^{(m)})_h$ exist for all sufficiently large m, then A_h exists, and $(A^{(m)})_h \rightrightarrows A_h$ as $m \to \infty$.*

e) *If A_h exists and $p < \infty$, then $(A^*)_h$ exists and is equal to $(A_h)^*$.*

Proof. a)–d) immediately follow from Definition 3.3 and Corollary 1.70.

e) First note that, for every $k \in \mathbb{N}$ and $\alpha \in \mathbb{Z}^n$, P_k^* and V_α^* can be identified with P_k and $V_{-\alpha}$ on $E^* = \ell^q(\mathbb{Z}^n, \mathbf{X}^*)$ with $1/p + 1/q = 1$, respectively. Then the fact that $\|P_k(V_{-h_m} A^* V_{h_m} - (A_h)^*)\| = \|((V_{-h_m} A V_{h_m} - A_h) P_k)^*\| = \|(V_{-h_m} A V_{h_m} - A_h) P_k\| \to 0$ as $m \to \infty$ for every $k \in \mathbb{N}$, together with its symmetric analogue, proves $V_{-h_m} A^* V_{h_m} \xrightarrow{\mathcal{P}} (A_h)^*$ as $m \to \infty$. □

Lemma 3.5. *Let $A \in L(E, \mathcal{P})$, and let $h \subset \mathbb{Z}^n$ lead to a limit operator of A. Then A_h cannot have a higher information flow than A; that is $f_{A_h}(d) \leq f_A(d)$ for all $d \in \mathbb{N}_0$, where the information flow $f_A(d)$ is defined by (1.30).*

Proof. Let $d \in \mathbb{N}_0$ be arbitrary, and take two sets $U, V \subset \mathbb{Z}^n$ with $\mathrm{dist}\,(U, V) \geq d$. Then, from $P_V A_h P_U = \mathcal{P}\text{-}\lim_{m \to \infty} P_V V_{-h_m} A V_{h_m} P_U$ and Corollary 1.70 a), we get

$$
\begin{aligned}
\|P_V A_h P_U\| &\leq \sup_m \|P_V V_{-h_m} A V_{h_m} P_U\| = \sup_m \|V_{-h_m} P_{h_m + V} A P_{h_m + U} V_{h_m}\| \\
&= \sup_m \|P_{h_m + V} A P_{h_m + U}\| \leq f_A(d)
\end{aligned}
$$

since V_α is an isometry for all $\alpha \in \mathbb{Z}^n$, and $\mathrm{dist}\,(h_m + U, h_m + V) = \mathrm{dist}\,(U, V) \geq d$ for all $m \in \mathbb{N}$. □

Proposition 3.6. *Take an $A \in L(E)$ and a sequence $h \subset \mathbb{Z}^n$ that leads to a limit operator of A.*

a) *If $A \in \mathcal{S}$, then also $A_h \in \mathcal{S}$, and $(A_h)^{\triangleleft} = (A^{\triangleleft})_h =: A_h^{\triangleleft}$.*

b) *If $A \in L(E, \mathcal{P})$, then also $A_h \in L(E, \mathcal{P})$.*

c) *If $A \in K(E, \mathcal{P})$, then $A_h = 0$ for any sequence h tending to infinity.*

d) *If A is a generalized multiplication operator, then also A_h is one.*

e) *If $A \in \mathrm{BO}$, then also $A_h \in \mathrm{BO}$.*

f) *If $A \in \mathrm{BDO}^p$, then also $A_h \in \mathrm{BDO}^p$.*

Proof. a) Repeat the proof of Proposition 3.4 e) with $A^{\triangleleft} \in L(E^{\triangleleft})$ in place of A.

b) follows from $\{V_\alpha : \alpha \in \mathbb{Z}^n\} \subset L(E, \mathcal{P})$ and Proposition 1.68.

c) For every $k \in \mathbb{N}$, put $U_k := \{-k, \ldots, k\}^n$. From $V_\alpha P_k = V_\alpha P_{U_k} = P_{\alpha + U_k} V_\alpha$ for all $\alpha \in \mathbb{Z}^n$, $k \in \mathbb{N}$ and from V_α being an isometry for every $\alpha \in \mathbb{Z}^n$, we conclude that, for a sequence $h = (h_m)$ tending to infinity and an operator $A \in L(E)$, the limit operator A_h equals 0 if and only if both

$$\|P_k(V_{-h_m} A V_{h_m} - 0)\| = \|V_{-h_m} P_{h_m + U_k} A V_{h_m}\| = \|P_{h_m + U_k} A\| \le \|Q_{h_m - k - 1} A\|,$$

$$\|(V_{-h_m} A V_{h_m} - 0)P_k\| = \|V_{-h_m} A P_{h_m + U_k} V_{h_m}\| = \|A P_{h_m + U_k}\| \le \|A Q_{h_m - k - 1}\|$$

tend to zero as $m \to \infty$ for every fixed $k \in \mathbb{N}$. But if $A \in K(E, \mathcal{P})$ and $h_m \to \infty$, this is clearly the case.

d–f) follow from Lemma 3.5 and the characterizations of the respective operator classes in terms of the function f_A in Subsection 1.3.7. $\qquad\square$

For the study of the invertibility at infinity of A, a single limit operator A_h in general cannot tell the whole story. That's why we collect **all** "samples" A_h that we can get from A.

Definition 3.7. *The set of all limit operators of $A \in L(E)$ is denoted by $\sigma^{\mathrm{op}}(A)$, and we refer to it as the operator spectrum of A.*

The operator spectrum includes all limit operators A_h of A, regardless of the direction in which h tends to infinity. But sometimes this direction is of importance, and so we will split $\sigma^{\mathrm{op}}(A)$ into many sets – the so called *local operator spectra*.

Definition 3.8. *For every direction $s \in S^{n-1}$, the local operator spectrum $\sigma_s^{\mathrm{op}}(A)$ is defined as the set of all limit operators A_h with $h = (h_m) \subset \mathbb{Z}^n$ and $h_m \to \infty_s$.*

For $n = 1$, we will abbreviate $\sigma_{-1}^{\mathrm{op}}(A)$ and $\sigma_{+1}^{\mathrm{op}}(A)$ by $\sigma_-^{\mathrm{op}}(A)$ and $\sigma_+^{\mathrm{op}}(A)$, respectively. The following result is certainly not surprising.

Proposition 3.9. *For every operator $A \in L(E)$, the identity*

$$\sigma^{\mathrm{op}}(A) = \bigcup_{s \in S^{n-1}} \sigma_s^{\mathrm{op}}(A)$$

holds.

Proof. Clearly, every local operator spectrum is contained in the operator spectrum $\sigma^{\mathrm{op}}(A)$. For the reverse inclusion, take some $A_h \in \sigma^{\mathrm{op}}(A)$. The sequence $h_m/|h_m|_2$ need not converge to a point s in S^{n-1}; but since the unit sphere S^{n-1} is compact, there is a subsequence g of h that has this property, and hence g tends to infinity in some direction $s \in S^{n-1}$. But then $A_h = A_g \in \sigma_s^{\mathrm{op}}(A)$.   □

As so often, we have to make sure that we have picked a sufficient amount of samples before we can make a reliable statement. Remember Example 3.1 and our little discussion afterwards. In order to fully characterize the invertibility of the object $a \in \mathbf{B}$, we needed a sufficient family of unital homomorphisms; this is why we had to look at $a \in \mathbf{B}$ everywhere – well, at least in a dense subset – in $[0, 1]$. The same problem will occur in connection with limit operators. In order to actually tell the whole story about A being invertible at infinity or not, its operator spectrum $\sigma^{\mathrm{op}}(A)$ has to contain sufficiently many elements in the following sense.

Definition 3.10. *We will say that $A \in L(E)$ is a rich operator if every sequence $h \subset \mathbb{Z}^n$, that tends to infinity, possesses a subsequence g which leads to a limit operator of A. The set of all rich operators $A \in L(E)$ is denoted by[2] $L_\$(E)$.*

As a rule, if we want to say anything about the operator A that is based on knowledge about its operator spectrum only, then A should better be rich in order to allow such a statement. In this connection we introduce the abbreviations

$$
\begin{aligned}
L_\$(E, \mathcal{P}) &:= L(E, \mathcal{P}) \cap L_\$(E), \\
\mathrm{BO}_\$ &:= \mathrm{BO} \cap L_\$(E), \\
\mathrm{BDO}_\$^p &:= \mathrm{BDO}^p \cap L_\$(E) \qquad \text{and} \\
\mathrm{BDO}_{S,\$}^p &:= \mathrm{BDO}_S^p \cap L_\$(E),
\end{aligned}
$$

where

$$
\mathrm{BDO}_S^p := \begin{cases} \mathrm{BDO}^p, & p < \infty, \\ \mathrm{BDO}^\infty \cap S, & p = \infty. \end{cases}
$$

Proposition 3.11.

a) $L_\$(E, \mathcal{P})$, $\mathrm{BDO}_\$^p$ *and* $\mathrm{BDO}_{S,\$}^p$ *are Banach subalgebras of $L(E)$.*

b) *The algebra $\mathrm{BO}_\$$ is dense in $\mathrm{BDO}_\$^p$.*

Proof. a) Let $A, B \in L_\$(E, \mathcal{P})$, and take an arbitrary sequence $h = (h_m) \subset \mathbb{Z}^n$ tending to infinity. Since A is rich, there is a subsequence g of h which leads to a limit operator of A. But since also B is rich, we can once more choose a subsequence f of g for which B_f exists, and by Proposition 3.4 b) and c), f leads to a limit operator of $A + B$ and of AB.

The closedness of $L_\$(E, \mathcal{P})$ is verified by an analogous construction: Take a sequence $(A^{(m)})$ in $L_\$(E, \mathcal{P})$ with $A^{(m)} \rightrightarrows A$ as $m \to \infty$ and an arbitrary sequence $h \subset \mathbb{Z}^n$ tending to infinity. From h we choose a subsequence $h^{(1)}$ leading to a limit

---

[2] Feel free to replace the dollar sign by any other currency of your choice.

operator of a $A^{(1)}$. From $h^{(1)}$ we choose a subsequence $h^{(2)}$ leading to a limit operator of $A^{(2)}$, and so on. Putting $g := (h_m^{(m)})$, and finally applying Proposition 3.4 d), we get $A \in L_{\$}(E, \mathcal{P})$.

The claim for $\mathrm{BDO}_{\$}^p$ follows from the fact that BDO^p is a Banach subalgebra of $L(E, \mathcal{P})$, and for $\mathrm{BDO}_{\mathcal{S},\$}^p$ it suffices to recall Proposition 1.10.

b) As the intersection of a (non-closed) algebra and a Banach algebra, $\mathrm{BO}_{\$} = \mathrm{BO} \cap L_{\$}(E, \mathcal{P})$ is a (non-closed) algebra. That it is dense in $\mathrm{BDO}_{\$}^p$ goes back to [50, Proposition 2.9] (also see [70, Theorem 2.1.18]). The proof uses details of the proof of Theorem 1.42.   □

We will say that $A \in L(E)$ is *translation-* or *shift invariant* if A commutes with every shift operator V_α with $\alpha \in \mathbb{Z}^n$. If $A \in L(E)$ is shift invariant, then the sequence (3.5) is constant, and A_h exists and is equal to the operator A itself for every sequence $h \subset \mathbb{Z}^n$ that tends to infinity.

3.2   Limit Operators Versus Invertibility at Infinity

The aim of this section is to discuss and prove the connection between an operator's invertibility at infinity and its operator spectrum. Clearly, invertibility at infinity is a property that only depends on an operator's behaviour in a neighbourhood of infinity. For a rich operator, it turns out that all necessary information about this behaviour is accurately stored in the operator spectrum $\sigma^{\mathrm{op}}(A)$.

3.2.1   Some Questions Around Theorem 1

For operators $A \in \mathrm{BDO}_{\mathcal{S},\$}^p$, we are able to prove Theorem 1 from the introduction of this book, saying that A is invertible at infinity if and only if its operator spectrum $\sigma^{\mathrm{op}}(A)$ is uniformly invertible. This is the central result which basically justifies the study of limit operators – the 'aorta' of the whole business. Before we come to its proof, we will discuss its practicability by looking at the following questions.

Q1   *I have an operator $A \in L(E)$.*
      How can I know whether it is in $\mathrm{BDO}_{\mathcal{S},\$}^p$ or not?

Q2   *Theorem 1 tells me I have to look at the limit operators of A.*
      How do I compute these?

Q3   *Are the limit operators A_h always of reasonably "simpler" structure than the operator A itself?*

We do not know an efficient algorithm that answers question **Q1** in general. We have certainly found some characterizations for the property $A \in \mathrm{BDO}^p$ in

Subsection 1.3.6, and we have at least Proposition 1.9 for the existence of a pre-adjoint, but how can we know whether A is rich or not? By Proposition 3.11, BDO$_{\mathcal{S},\$}^{p}$ is a Banach algebra. So we can at least try to decompose our operator A into some basic components from which it is assembled via addition, composition and passing to the norm-limit and find out whether or not these components are rich. If they are, then A is rich as well; if they are not, then A is probably not.

The natural candidates for these basic components are shift operators and generalized multiplication operators, which are the atoms of every band-dominated operator as we pointed out in Remark 1.38. In this connection recall that we referred to the symbols of the generalized multiplication operators arising from the decomposition of a band operator as its coefficients, and that we say that $A \in \text{BDO}^{p}$ has coefficients in a set F if it is the norm limit of a sequence of band operators with coefficients in F. In Chapter 4 we will moreover study a class of integral operators $A \in L(L^{p})$ which are composed by convolution operators (remember Example 1.45) and usual multiplication operators on L^{p}. In this case, we will clearly think of these two classes of operators as the atoms of A. In either case, the problem can be reduced to the study of generalized and usual multiplication operators since the other components, namely shift operators and convolution operators, are shift-invariant, and hence these are rich operators. The remaining question, whether or not a (generalized) multiplication operator is rich, is studied in Section 3.4.

Concerning question $\mathbf{Q2}$, we make use of the facts we proved in Proposition 3.4. So computing a limit operator of A can be done by decomposing A into its atoms, computing their limit operators, and puzzling these together again as A was assembled by its components. Again, limit operators of a shift-invariant operator are equal to the operator itself, and hence the problem can be reduced to the computation of the limit operators of (generalized) multiplication operators. Also this issue is discussed in Section 3.4.

Question $\mathbf{Q3}$ is also a very tough one because it is not clear at all what it means that an operator is "simpler" than another. We will address some aspects of the apparent simplification $A \mapsto A_{h}$ in Section 3.8.

3.2.2   Proof of Theorem 1

In this subsection we will prove Theorem 1. We divide the proof in some propositions and lemmas. The reader who is not interested in this proof might simply want to skip the rest of this section and continue at Section 3.3 on page 86.

For the rest of this section we will fix the following set of continuous functions. Also remember that we associated a generalized multiplication operator $\hat{M}_{f}$ with every bounded and continuous function f on $\mathbb{R}^{n}$ by (1.25).

By $\vartheta : \mathbb{R} \to [0,1]$ we denote the continuous function

$$\vartheta(x) \; := \; \begin{cases} 3x+2 & \text{if } x \in [-2/3,\, -1/3), \\ 1 & \text{if } x \in [-1/3,\, 1/3], \\ -3x+2 & \text{if } x \in (1/3,\, 2/3], \\ 0 & \text{otherwise,} \end{cases} \qquad (3.6)$$

with support in $[-2/3,\, 2/3]$. Let $\omega : \mathbb{R}^n \to [0,1]$ be the continuous function with support in $[-2/3,\, 2/3]^n$ which is given by

$$\omega(x_1, \ldots, x_n) \; := \; \sqrt{\vartheta(x_1) \cdots \vartheta(x_n)}, \qquad (x_1, \ldots, x_n) \in \mathbb{R}^n.$$

For every $\alpha \in \mathbb{Z}^n$, we put

$$\omega_\alpha(x) \; := \; \omega(x - \alpha), \qquad x \in \mathbb{R}^n.$$

Note that ω is chosen such that

$$\sum_{\alpha \in \mathbb{Z}^n} \omega_\alpha^2(x) \; = \; 1, \qquad x \in \mathbb{R}^n$$

holds. In this sense we say that $\{\omega_\alpha^2\}_{\alpha \in \mathbb{Z}^n}$ is a partition of unity. Also note that, for every $x \in \mathbb{R}^n$, the latter sum is actually a finite sum. For every $R > 0$, we moreover define the function

$$\omega_{\alpha,R}(x) \; := \; \omega_\alpha(x/R) \; = \; \omega(x/R - \alpha), \qquad x \in \mathbb{R}^n.$$

Obviously, $\{\omega_{\alpha,R}^2\}_{\alpha \in \mathbb{Z}^n}$ is a partition of unity as well for every $R > 0$.

Finally, let $\psi : \mathbb{R}^n \to [0,1]$ be a continuous function supported in $[-4/5, 4/5]^n$ which is equal to 1 on $[-3/4,\, 3/4]^n$ (and hence on the whole support of ω). In analogy to ω_α and $\omega_{\alpha,R}$, we define ψ_α and $\psi_{\alpha,R}$, respectively.

Proposition 3.12. *If $A \in \mathrm{BDO}_\mathcal{S}^p$ is invertible at infinity, then its operator spectrum $\sigma^{\mathrm{op}}(A)$ is uniformly invertible.*

Proof. We demonstrate the proof for the case $p = \infty$, where we make use of the existence of the pre-dual $E^\lhd$ and the pre-adjoint operator $A^\lhd$. For $p < \infty$, the proof is almost completely analogously, with $E^\lhd$ and $A^\lhd$ replaced by E^* and A^*, respectively.

Let $E = \ell^\infty(\mathbb{Z}^n, \mathbf{X})$, and suppose $A \in \mathrm{BDO}_\mathcal{S}^p$ is invertible at infinity. Then there is a $B \in L(E)$ such that $I = BA + T_1$ and $I = AB + T_2$ with $T_1, T_2 \in K(E, \mathcal{P})$. For arbitrary elements $u \in E$ and $v \in E^*$, by passing to adjoints in the second equation, this yields to

$$\|u\|_E \; \leq \; \|B\| \|Au\|_\infty + \|T_1 u\|_E$$
$$\|v\|_{E^*} \; \leq \; \|B^*\| \|A^* v\|_{E^*} + \|T_2^* v\|_{E^*}.$$

Restricting v to $E^\lhd \subset E^*$ and putting $s := \max(\|B\|, 1)$, we get that the inequalities

$$\|u\|_E \leq s(\|Au\|_E + \|T_1 u\|_E) \tag{3.7}$$
$$\|v\|_{E^\lhd} \leq s(\|A^\lhd v\|_{E^\lhd} + \|T_2^* v\|_{E^*}) \tag{3.8}$$

hold for all $u \in E$ and $v \in E^\lhd$.

Let $h = (h_m) \subset \mathbb{Z}^n$ be an arbitrary sequence tending to infinity and leading to a limit operator A_h of A. We will show that A_h is invertible.

Since V_α is an isometry for every $\alpha \in \mathbb{Z}^n$, we get from (3.7) that

$$\|u\|_E = \|V_{h_m} u\|_E \leq s(\|AV_{h_m} u\|_E + \|T_1 V_{h_m} u\|_E)$$
$$= s(\|V_{-h_m} AV_{h_m} u\|_E + \|V_{-h_m} T_1 V_{h_m} u\|_E), \quad u \in E.$$

Remember the functions $\psi_{\alpha, R}$, defined above for all $\alpha \in \mathbb{Z}^n$ and $R > 0$. Replacing u in the last inequality by $\hat{M}_{\psi_{0,R}} u$ and doing all the same steps with inequality (3.8), we get that

$$\|\hat{M}_{\psi_{0,R}} u\|_E \leq s(\|V_{-h_m} AV_{h_m} \hat{M}_{\psi_{0,R}} u\|_E + \|V_{-h_m} T_1 V_{h_m} \hat{M}_{\psi_{0,R}} u\|_E),$$
$$\|\hat{M}_{\psi_{0,R}} v\|_{E^\lhd} \leq s(\|V_{-h_m} A^\lhd V_{h_m} \hat{M}_{\psi_{0,R}} v\|_{E^\lhd} + \|V_{-h_m} T_2^* V_{h_m} \hat{M}_{\psi_{0,R}} v\|_{E^*})$$

are true for all $R > 0$, all $u \in E$ and all $v \in E^\lhd$.

From $(V_{-h_m} AV_{h_m} - A_h)\hat{M}_{\psi_{0,R}} \rightrightarrows 0$, Proposition 3.6b), $V_{-h_m} T_1 V_{h_m} \hat{M}_{\psi_{0,R}} \rightrightarrows 0$ and $\|V_{-h_m} T_2^* V_{h_m} \hat{M}_{\psi_{0,R}}\| = \|(\hat{M}_{\psi_{0,R}} V_{-h_m} T_2 V_{h_m})^*\| = \|\hat{M}_{\psi_{0,R}} V_{-h_m} T_2 V_{h_m}\|$ going to zero as $m \to \infty$, we conclude, for all $R > 0$, all $u \in E$ and all $v \in E^\lhd$,

$$\|\hat{M}_{\psi_{0,R}} u\|_E \leq s\|A_h \hat{M}_{\psi_{0,R}} u\|_E \leq s(\|\hat{M}_{\psi_{0,R}} A_h u\|_E + \|[A_h, \hat{M}_{\psi_{0,R}}] u\|_E),$$
$$\|\hat{M}_{\psi_{0,R}} v\|_{E^\lhd} \leq s\|A_h^\lhd \hat{M}_{\psi_{0,R}} v\|_{E^\lhd} \leq s(\|\hat{M}_{\psi_{0,R}} A_h^\lhd v\|_{E^\lhd} + \|[A_h^\lhd, \hat{M}_{\psi_{0,R}}] v\|_{E^\lhd}).$$

Since $A_h \in \mathrm{BDO}^p$ and $[A_h^\lhd, \hat{M}_{\psi_{0,R}}] = [\hat{M}_{\psi_{0,R}}, A_h]^\lhd$ holds, letting $R \to \infty$ leads to

$$\|u\|_E \leq s\|A_h u\|_E, \quad u \in E,$$
$$\|v\|_{E^\lhd} \leq s\|A_h^\lhd v\|_{E^\lhd}, \quad v \in E^\lhd.$$

So both A_h and $A_h^\lhd$ are bounded below with a lower norm not less than $1/s$. From Lemma 2.35 we conclude that $A_h^\lhd$ is invertible with its inverse being bounded by s. Consequently, also its adjoint A_h is invertible and also $\|A_h^{-1}\| \leq s$, where s does not depend on the sequence h. $\qquad\square$

The rest of this subsection is devoted to the proof of the sufficiency part of Theorem 1. Note that the condition $A \in \mathcal{S}$ (for $p = \infty$) was only needed for the necessity part, whereas the condition that A is a rich operator is only needed in the sufficiency part.

Proposition 3.13. *Let $A \in \mathrm{BDO}^p$, and suppose a limit operator A_h of A is invertible. Then, for every function $\xi \in L^\infty$ with bounded support, there exists a $m_0 \in \mathbb{N}$ such that, for every $m > m_0$, there are operators $B_m, C_m \in \mathrm{BDO}^p$ with $\|B_m\|, \|C_m\| < 2\|(A_h)^{-1}\|$ and*

$$B_m A \hat{M}_{V_{h_m} \xi} \;=\; \hat{M}_{V_{h_m} \xi} \;=\; \hat{M}_{V_{h_m} \xi} A C_m.$$

Proof. The proof of this proposition is very straightforward. It can be found in [67, Proposition 15], [70, Proposition 2.2.4], [47, Proposition 3.3] or [50, Proposition 3.2]. $\qquad\square$

Lemma 3.14. *If $\{A_\alpha\}_{\alpha \in \mathbb{Z}^n} \subset L(E, \mathcal{P})$ is bounded, then, for every $R > 0$, the operator*

$$A_R \;:=\; \sum_{\alpha \in \mathbb{Z}^n} \hat{M}_{\omega_{\alpha,R}} A_\alpha \hat{M}_{\psi_{\alpha,R}}$$

is well-defined, and $\|A_R\| \le 2^n \sup_{\alpha \in \mathbb{Z}^n} \|A_\alpha\|$ holds. For $p < \infty$, the series is $\mathcal{P}$-convergent, and for $p = \infty$, the operator can still be defined pointwise $(A_R u)(x)$ for every $u \in E$ and every $x \in \mathbb{Z}^n$.

Proof. For a proof for $p < \infty$, see [67, Proposition 13] or [70, Proposition 2.2.2], and for $p = \infty$ this follows from [47, Lemma 2.6] or [50, Lemma 1.8]. $\qquad\square$

Proposition 3.15. *Let A be some operator in BDO^p. If there is an $M > 0$ such that, for every $R > 0$, there exists a $\rho(R)$ such that, for every $\alpha \in \mathbb{Z}^n$ with $|\alpha| > \rho(R)$, there are two operators $B_{\alpha,R}, C_{\alpha,R} \in L(E, \mathcal{P})$ with $\|B_{\alpha,R}\|, \|C_{\alpha,R}\| < M$ and*

$$B_{\alpha,R} A \hat{M}_{\omega_{\alpha,R}} \;=\; \hat{M}_{\omega_{\alpha,R}} \;=\; \hat{M}_{\omega_{\alpha,R}} A C_{\alpha,R}, \qquad (3.9)$$

then A is invertible at infinity.

Proof. If $A \in \mathrm{BO}$, then it can be verified that, for a sufficiently large $R > 0$, the operators

$$B_R := \sum_{|\alpha| > \rho(R)} \hat{M}_{\omega_{\alpha,R}} B_{\alpha,R} \hat{M}_{\omega_{\alpha,R}} \qquad \text{and} \qquad C_R := \sum_{|\alpha| > \rho(R)} \hat{M}_{\omega_{\alpha,R}} C_{\alpha,R} \hat{M}_{\omega_{\alpha,R}}$$

are subject to

$$B_R A = I - \sum_{|\alpha| \le \rho(R)} \hat{M}^2_{\omega_{\alpha,R}} + T_1 \qquad \text{and} \qquad A C_R = I - \sum_{|\alpha| \le \rho(R)} \hat{M}^2_{\omega_{\alpha,R}} + T_2. \quad (3.10)$$

where $T_1, T_2 \in L(E, \mathcal{P})$ and $\|T_1\|, \|T_2\| < 1/2$. The operators B_R and C_R are defined by Lemma 3.14. Then $I + T_1$ and $I + T_2$ are invertible, their inverses are in $L(E, \mathcal{P})$ again, and multiplying the first equality in (3.10) from the left by $(I + T_1)^{-1}$ and the second one from the right by $(I + T_2)^{-1}$, shows that A is $\mathcal{P}$-Fredholm, and hence invertible at infinity.

Using this fact and Lemma 1.3, the result can be extended to all $A \in \mathrm{BDO}^p$. For further details of the proof, see [70, Proposition 2.2.3]. $\qquad\square$

Now we are ready for the proof of the sufficiency part of Theorem 1, which, together with Proposition 3.12, completes the proof of Theorem 1.

Proposition 3.16. *If $A \in \mathrm{BDO}^p_\$$, and its operator spectrum $\sigma^{\mathrm{op}}(A)$ is uniformly invertible, then A is invertible at infinity.*

Proof. Let $\sigma^{\mathrm{op}}(A)$ be uniformly invertible, and suppose A were not invertible at infinity. Then, by Proposition 3.15, for every $M > 0$, and hence, for

$$M := 2 \sup_{A_h \in \sigma^{\mathrm{op}}(A)} \|A_h^{-1}\|,$$

there is an $R > 0$ such that, for all $\rho > 0$, there exists an $\alpha \in \mathbb{Z}^n$ with $|\alpha| > \rho$ such that

$$B A \hat{M}_{\omega_{\alpha,R}} \neq \hat{M}_{\omega_{\alpha,R}}$$

for all operators $B \in L(E,\mathcal{P})$ with $\|B\| < M$. Consequently, there is a sequence $(\alpha_k) \subset \mathbb{Z}^n$ tending to infinity such that, for no $k \in \mathbb{N}$, there is an operator $B_k \in L(E,\mathcal{P})$ with $\|B_k\| < M$ for which

$$B_k A \hat{M}_{\omega_{\alpha_k},R} = \hat{M}_{\omega_{\alpha_k},R} \tag{3.11}$$

holds.

Since A is rich, there is a subsequence g of (α_k) such that A_g exists. But $A_g \in \sigma^{\mathrm{op}}(A)$ is invertible by our premise. And so Proposition 3.13 (putting $\xi := \omega_{0,R}$) says that, for infinitely many $k \in \mathbb{N}$, there is an operator $B_k \in \mathrm{BDO}^p \subset L(E,\mathcal{P})$ with $\|B_k\| < M$ that is subject to equation (3.11). This is clearly a contradiction.   □

3.3   Fredholmness, Lower Norms and Pseudospectra

3.3.1   Fredholmness vs. Limit Operators

From Subsection 2.3 we know that Fredholmness is closely related, sometimes even coincident, with invertibility at infinity; the latter can be studied by looking at limit operators. Here is a first simple result along these lines.

Proposition 3.17. *Let $E = \ell^p(\mathbb{Z}^n, \mathbf{X})$ with $p \in [1,\infty]$ and a Banach space $\mathbf{X}$.*

a) *If $1 < p < \infty$ and $A \in \mathrm{BDO}^p$ is Fredholm, then all limit operators of A are invertible, and their inverses are uniformly bounded. Moreover,*

$$\bigcup_{B \in \sigma^{\mathrm{op}}(A)} \mathrm{sp}\, B \subset \mathrm{sp}_{\mathrm{ess}}\, A. \tag{3.12}$$

b) *Now suppose $A \in \mathrm{BDO}^p_\$$, and either A is the sum of an invertible and a locally compact operator or $\dim \mathbf{X} < \infty$. Then, if $\sigma^{\mathrm{op}}(A)$ is uniformly invertible, A is Fredholm.*

Proof. a) From $1 < p < \infty$ and Figure 4 we get that A is invertible at infinity if it is Fredholm. The uniform invertibility of $\sigma^{\mathrm{op}}(A)$ follows now from Proposition 3.12, where the condition $A \in \mathcal{S}$ is redundant if $p \neq \infty$.

The inclusion (3.12) follows analogously: Suppose $B - \lambda I$ is not invertible for $\lambda \in \mathbb{C}$ and some limit operator B of A. From $B - \lambda I \in \sigma^{\mathrm{op}}(A - \lambda I)$ and Proposition 3.12 we get that $A - \lambda I$ is not invertible at infinity and hence not Fredholm by Figure 4 if $1 < p < \infty$.

b) This is immediate from Propositions 3.16, 2.15 and Figure 4. $\qquad\square$

For the reverse implication of (3.12) – as in (3.52), for example – we need that the elementwise invertibility of $\sigma^{\mathrm{op}}(A)$ implies its uniform invertibility. For details about this, see Section 3.9, in particular Subsection 3.9.4.

For the remainder of this section, suppose $n = 1$, $1 < p < \infty$ and $\mathbf{X} = \mathbb{C}$. Consequently, Fredholmness coincides with invertibility at infinity, and Theorem 1 says that $A \in \mathrm{BDO}_{\$}^p$ is Fredholm if and only if its operator spectrum is uniformly invertible. Interestingly, even the Fredholm index of A can be restored from its operator spectrum $\sigma^{\mathrm{op}}(A)$!

We abbreviate the projector to the positive half axis and its complementary projector by $P := P_{\mathbb{N}}$ and $Q := I - P$, respectively. Now let us suppose that $A \in \mathrm{BDO}^p = \mathrm{BDO}_{\$}^p$ (see Corollary 3.24) on $E = \ell^p = \ell^p(\mathbb{Z}, \mathbb{C})$. Via the evident formula $E = \mathrm{im}\,P \dotplus \mathrm{im}\,Q$, we can decompose A into four operators:

$$A = \left(\begin{array}{c|c} QAQ & QAP \\ \hline PAQ & PAP \end{array}\right) = \left(\begin{array}{cc} \blacksquare & \blacksquare \\ \blacksquare & \blacksquare \end{array}\right).$$

If $A \in \mathrm{BO}$, then the two blocks PAQ and QAP are finite-rank operators, whence they are compact if A is band-dominated. Consequently,

$$A = (P + Q)A(P + Q) = PAP + PAQ + QAP + QAQ$$

is Fredholm if and only if

$$A - PAQ - QAP = PAP + QAQ = \left(\begin{array}{cc} \blacksquare & \\ & \blacksquare \end{array}\right) \tag{3.13}$$

is Fredholm, where the two Fredholm indices coincide in this case. For the study
of the operator (3.13), we put

$$A_+ := PAP + Q \quad \text{and} \quad A_- := QAQ + P, \tag{3.14}$$

which gives us that (3.13) equals the product $A_+A_- = A_-A_+$.

From this equality it follows that (3.13), and hence A, is Fredholm if and
only if A_+ and A_- are Fredholm, and

$$\text{ind}\, A = \text{ind}\, A_+ + \text{ind}\, A_- \tag{3.15}$$

holds. Clearly,

$$\sigma^{\text{op}}(A_+) = \sigma_+^{\text{op}}(A) \cup \{I\} \quad \text{and} \quad \sigma^{\text{op}}(A_-) = \sigma_-^{\text{op}}(A) \cup \{I\},$$

whence the Fredholmness of $A_\pm$ is determined by the local operator spectrum
$\sigma_\pm^{\text{op}}(A)$, respectively. But also the index of $A_\pm$ is hidden, in an astonishingly simple
way, in $\sigma_\pm^{\text{op}}(A)$, respectively. Indeed, if we call

$$\text{ind}^+ A := \text{ind}\, A_+ = \text{ind}\, (PAP + Q)$$

and

$$\text{ind}^- A := \text{ind}\, A_- = \text{ind}\, (QAQ + P)$$

the *plus-* and the *minus-index* of the band-dominated operator A, then the following result holds.

Proposition 3.18. *Let $A \in \text{BDO}^p$ be Fredholm. Then all operators in $\sigma_+^{\text{op}}(A)$ have
the same plus-index, and this number coincides with the plus-index of A. Analo-
gously, all operators in $\sigma_-^{\text{op}}(A)$ have the same minus-index, and this number coincides with the minus-index of A.*

This remarkable result was derived in [66] via computations of the K-group
of the C^*-algebra BDO^2 for $p = 2$, and it was generalized to $1 < p < \infty$ in [74].
It should be mentioned that ROCH's results in [74] easily extend to operators in
BDO^1 and BDO^∞ that belong to the Wiener algebra $\mathcal{W}$.

Corollary 3.19. *Let $A \in \text{BDO}^p$ be Fredholm. Then, for any two limit operators
$B \in \sigma_+^{\text{op}}(A)$ and $C \in \sigma_-^{\text{op}}(A)$ of A, the identities*

$$\begin{aligned}
\text{ind}^+ A &= \text{ind}^+ B = -\text{ind}^- B \\
\text{ind}^- A &= \text{ind}^- C = -\text{ind}^+ C \\
\text{ind}\, A &= \text{ind}^+ B + \text{ind}^- C = \text{ind}\, (PBP + Q) + \text{ind}\, (QCQ + P)
\end{aligned}$$

hold.

Proof. If A is Fredholm, then B and C are invertible by Theorem 1, whence $\text{ind}\, B$
and $\text{ind}\, C$ are both equal to zero. The rest is immediate from Proposition 3.18 and
(3.15).   □

3.3.2 Pseudospectra vs. Limit Operators

Besides close connections between the essential spectrum of a band-dominated operator A and the spectra of its limit operators, as indicated in the previous subsection, one can also say something about the pseudospectrum [85] of A in terms of the pseudospectra of its limit operators.

Again, put $E = \ell^p(\mathbb{Z}^n, \mathbf{X})$ with $p \in [1, \infty]$, $n \in \mathbb{N}$ and an arbitrary Banach space $\mathbf{X}$. As an auxiliary result, we first say something about lower norms. Therefore recall Definition 2.31 and equation (1.13).

Lemma 3.20.   a) *Let $A \in L(E)$ and $B \in \sigma^{\mathrm{op}}(A)$. Then $\|Bu\| \geq \nu(A)\,\|u\|$ for all $u \in E_0$. In particular, $\nu(B) \geq \nu(A)$ if $E_0 = E$; that is if $p < \infty$.*

b) *If $A \in L(E, \mathcal{P})$ and $B \in \sigma^{\mathrm{op}}(A)$ is invertible, then $\nu(B) \geq \nu(A)$.*

Proof. a) If $B \in \sigma^{\mathrm{op}}(A)$, then $B = A_h$ for some sequence $h = (h_m) \subset \mathbb{Z}^n$ tending to infinity. For $k \in \mathbb{N}$ and every $u \in E_0$, we have that

$$\|V_{-h_m} A V_{h_m} P_k u\| = \|A V_{h_m} P_k u\| \geq \nu(A)\,\|V_{h_m} P_k u\| = \nu(A)\,\|P_k u\|.$$

Since $V_{-h_m} A V_{h_m} \overset{P}{\to} B$, taking the limit as $m \to \infty$ we get

$$\|B P_k u\| \geq \nu(A)\,\|P_k u\|.$$

As $u \in E_0$ we have $P_k u \to u$ as $k \to \infty$, so the result follows.

b) For $k, l \in \mathbb{N}$ and $u \in E$,

$$
\begin{aligned}
\|P_k B^{-1} u\| &\leq \|P_k B^{-1} Q_l u\| + \|P_k B^{-1} P_l u\| \\
&\leq \|P_k B^{-1} Q_l u\| + \|B^{-1} P_l u\|. \quad\quad (3.16)
\end{aligned}
$$

As $L(E, \mathcal{P})$ is inverse closed, by Proposition 1.46, we have that $B^{-1} \in L(E, \mathcal{P})$, so that $\|P_k B^{-1} Q_r\| \to 0$ and $\|Q_r B^{-1} P_l\| \to 0$ as $r \to \infty$, the latter implying that $B^{-1} P_l u \in E_0$. Thus from (3.16) and a) we have that

$$\nu(A)\,\|P_k B^{-1} u\| \leq \nu(A)\,\|P_k B^{-1} Q_l u\| + \|P_l u\|.$$

Taking the limit first as $l \to \infty$ and then as $k \to \infty$, we get that $\nu(A)\,\|B^{-1} u\| \leq \|u\|$. So we have shown that $\nu(A)\,\|v\| \leq \|Bv\|$, for all $v \in E$, as required. $\square$

For all $A \in L(E)$ and $\varepsilon > 0$, we define the ε-*pseudospectrum* of A by

$$\mathrm{sp}_\varepsilon A := \{\lambda \in \mathbb{C} : \|(\lambda I - A)^{-1}\| \geq 1/\varepsilon\},$$

where we again agree by putting $\|B^{-1}\| := \infty$ if B is not invertible. In this sense, if we moreover put $1/0 := \infty$, then $\mathrm{sp}_0 A = \mathrm{sp}\, A$ for all $A \in L(E)$.

Proposition 3.21. *For all $A \in \mathrm{BDO}_{\mathcal{S}}^p$ and all $\varepsilon > 0$, one has*

$$\bigcup_{B \in \sigma^{\mathrm{op}}(A)} \mathrm{sp}_\varepsilon B \subset \mathrm{sp}_\varepsilon A.$$

Proof. From Proposition 3.12 we know that, if $B \in \sigma^{op}(A)$ and $\lambda I - B$ is not invertible, then $\lambda I - A$ is not invertible. Now suppose $\lambda I - B$ is invertible. If $\lambda I - A$ is not invertible, then there is nothing to prove. If also $\lambda I - A$ is invertible, then, by Lemma 2.35 and Lemma 3.20 b), the latter applies since $B \in \sigma^{op}(A) \subset L(E, \mathcal{P})$ as $A \in \mathrm{BDO}^p \subset L(E, \mathcal{P})$, it follows that

$$\|(\lambda I - B)^{-1}\| = \frac{1}{\nu(\lambda I - B)} \leq \frac{1}{\nu(\lambda I - A)} = \|(\lambda I - A)^{-1}\|. \qquad \square$$

For a stronger result than Proposition 3.21 in the case when $L(E)$ is a C^*-algebra; that is when $p = 2$ and $\mathbf{X}$ is a Hilbert space, see [63] or Section 6.3 in [70].

3.4 Limit Operators of a Multiplication Operator

As discussed in Subsection 3.2.1, the practicability of Theorem 1 and of all its applications essentially rests on our knowledge about generalized and usual multiplication operators, more precisely, whether or not such an operator is rich, and what its limit operators look like.

3.4.1 Rich Functions

We start with generalized multiplication operators. So let $\mathbf{X}$ be a Banach space.

Proposition 3.22. *For* $b = (b_\alpha)_{\alpha \in \mathbb{Z}^n} \in \ell^\infty(\mathbb{Z}^n, L(\mathbf{X}))$, *the generalized multiplication operator* $\hat{M}_b$ *is rich if and only if the set* $\{b_\alpha\}_{\alpha \in \mathbb{Z}^n}$ *is relatively compact in* $L(\mathbf{X})$.

Proof. This is Theorem 2.1.16 in [70]. $\qquad \square$

Note that, if $\dim \mathbf{X} < \infty$, the boundedness of the set $\{b_\alpha\}_{\alpha \in \mathbb{Z}^n}$ automatically implies its relative compactness in $L(\mathbf{X})$.

Corollary 3.23. *If* $\dim \mathbf{X} < \infty$, *then every generalized multiplication operator on* $E = \ell^p(\mathbb{Z}^n, \mathbf{X})$ *is rich.*

Taking this together with Proposition 3.11, we even get the following.

Corollary 3.24. *If* $\dim \mathbf{X} < \infty$, *then* $\mathrm{BDO}^p_\$ = \mathrm{BDO}^p$. *So every band-dominated operator on* $E = \ell^p(\mathbb{Z}^n, \mathbf{X})$ *is rich.*

From this point we will mainly focus on multiplication operators on $E = L^p$. Remember that, by (1.2), $L^p = L^p(\mathbb{R}^n)$ can be identified with $\ell^p(\mathbb{Z}^n, \mathbf{X})$ where $\mathbf{X} = L^p(H)$ and $H = [0, 1)^n$. Via this identification, by (1.3), the multiplication operator M_b with symbol $b \in L^\infty$ corresponds to the generalized multiplication operator with symbol $(b_\alpha)_{\alpha \in \mathbb{Z}^n}$, where b_α is the operator of multiplication by the restriction of b to the cube $\alpha + H$, shifted to H.

From Proposition 3.22 we imply the following.

Corollary 3.25. *For $b \in L^\infty$, the multiplication operator M_b is rich if and only if the set $\{b|_{\alpha+H}\}_{\alpha \in \mathbb{Z}^n}$ is relatively compact in $L^\infty(H)$.*

Remark 3.26. Note that we wrote $b|_{\alpha+H}$ as a lazy abbreviation for the much clumsier notations $(V_{-\alpha}b)|_H$ or $V_{-\alpha}P_{\alpha+H}b$ in the previous corollary. Also note that the norm limit of a sequence of multiplication operators on $L^p(H)$, if it exists, is clearly a multiplication operator on $L^p(H)$ again. □

For multiplication operators on L^p, we have the following result which is some sort of a continuous analogue of Proposition 3.6 d). It says that the limit operator of a multiplication operator M_b is always a multiplication operator M_c where the symbol c is the $\mathcal{P}$-limit of a sequence of shifts of the symbol b.

Proposition 3.27. *Let $A = M_b$ with $b \in L^\infty$, and let $h = (h_m) \subset \mathbb{Z}^n$ tend to infinity. The following properties are equivalent, and, if they hold, then $A_h = M_c$.*

(i) *A_h exists.*

(ii) *There is a function $c \in L^\infty$ with*

$$\operatorname{ess\,sup}_{x \in U} |b(h_m + x) - c(x)| \ \to \ 0 \qquad as \qquad m \to \infty \qquad (3.17)$$

for all bounded and measurable sets $U \subset \mathbb{R}^n$.

(iii) *There is a function $c \in L^\infty$ such that (3.17) holds with $U = [-k,k]^n$ for every $k \in \mathbb{N}$.*

(iv) *The $\mathcal{P}$-limit $c := \mathcal{P}-\lim V_{-h_m}b$ exists for $m \to \infty$.*

Proof. Since $[-k,k]^n$ is bounded and measurable for every $k \in \mathbb{N}$, and since every bounded and measurable set $U \subset \mathbb{R}^n$ is contained in a set $[-k,k]^n$ with $k \in \mathbb{N}$, properties (ii) and (iii) are equivalent. Moreover, (iii) and (iv) are equivalent by Proposition 1.65 b).

(i)$\Rightarrow$(iii). If A_h exists, then from $M_{V_{-h_m}b} = V_{-h_m}M_b V_{h_m} \xrightarrow{\mathcal{P}} A_h$ as $m \to \infty$, we get that

$$M_{P_k V_{-h_m}b} = P_k M_{V_{-h_m}b} \rightrightarrows P_k A_h \qquad as \qquad m \to \infty$$

for every $k \in \mathbb{N}$. Consequently, $(P_k V_{-h_m}b)_{m=1}^\infty$ is a Cauchy sequence, and hence convergent, in $\operatorname{im} P_k = L^\infty([-k,k]^n)$ for every $k \in \mathbb{N}$. By Proposition 1.65 b), the sequence $(V_{-h_m}b)_{m=1}^\infty$ is $\mathcal{P}$-convergent in L^∞. Denote its $\mathcal{P}$-limit by $c \in L^\infty$. Then, for every $k \in \mathbb{N}$,

$$\operatorname{ess\,sup}_{x \in [-k,k]^n} |b(h_m + x) - c(x)| = \|P_k(V_{-h_m}b - c)\|_\infty \to 0 \qquad as \qquad m \to \infty,$$

which is (iii).

(iv)$\Rightarrow$(i). Suppose $P_k V_{-h_m} b \to P_k c$ as $m \to \infty$ for every $k \in \mathbb{N}$. So we have

$$P_k M_{V_{-h_m} b} \rightrightarrows P_k M_c \qquad \text{and} \qquad M_{V_{-h_m} b} P_k \rightrightarrows M_c P_k \qquad \text{as} \qquad m \to \infty$$

for every $k \in \mathbb{N}$, and hence $V_{-h_m} M_b V_{h_m} = M_{V_{-h_m} b} \xrightarrow{\mathcal{P}} M_c$ as $m \to \infty$, whence A_h exists and is equal to M_c. $\square$

The function $c = \mathcal{P}-\lim V_{-h_m} b$ from Proposition 3.27 will henceforth be denoted by $b^{(h)}$. So h leads to a limit operator of M_b if and only if $b^{(h)} = \mathcal{P}-\lim V_{-h_m} b$ exists in which case $(M_b)_h = M_{b^{(h)}}$.

Remark 3.28. a) We will sometimes write (3.17) in the form

$$b|_{h_m + U} \to c|_U \qquad \text{as} \qquad m \to \infty, \tag{3.18}$$

where the functions $b|_{h_m + U}$ are again identified with functions on U via V_{-h_m}.

b) Since the limit operator method grew up in the discrete case, the sequences $h = (h_m)$ were naturally restricted to $\mathbb{Z}^n$. In the continuous case we could easily drop this restriction and pass to $h = (h_m) \subset \mathbb{R}^n$. We will however resist this temptation and stay in the integers, which will result in some technical efforts in Subsections 3.4.2, 3.4.8 and 3.4.9. But afterwards, in Subsection 3.4.13, we will state the reason for doing so. $\square$

In addition to the cube $H = [0,1)^n$, we fix one more notation for a cube that we are using frequently in the following, which is $Q := [-1,1]^n$. Moreover, recall that, for $x = (x_1, \ldots, x_n) \in \mathbb{R}^n$, we denote by $[x] \in \mathbb{Z}^n$ the vector $([x_1], \ldots, [x_n])$ of the component-wise integer parts of x.

Definition 3.29. *For a function* $f \in L^\infty$, *a point* $x \in \mathbb{R}^n$ *and a bounded and measurable set* $U \subset \mathbb{R}^n$, *we define*

$$\mathrm{osc}_U(f) := \operatorname*{ess\,sup}_{u,v \in U} |f(u) - f(v)| \qquad \text{and} \qquad \mathrm{osc}_x(f) := \mathrm{osc}_{x+Q}(f),$$

the latter is referred to as local oscillation of f *at* x.

Lemma 3.30. *If* $b \in L^\infty$ *and* $h = (h_m) \subset \mathbb{Z}^n$ *leads to a limit operator* M_c *of* M_b, *then, for every bounded and measurable set* $U \subset \mathbb{R}^n$,

$$\mathrm{osc}_{h_m + U}(b) \to \mathrm{osc}_U(c) \qquad \text{as} \qquad m \to \infty.$$

Proof. Take an arbitrary $\varepsilon > 0$ and a bounded and measurable $U \subset \mathbb{R}^n$. By (3.17), there is some m_0 such that, for every $m > m_0$,

$$|b(h_m + x) - c(x)| < \frac{\varepsilon}{2}$$

holds for almost all $x \in U$. For almost all $u, v \in U$ we then have

$$\begin{aligned}
|b(h_m + u) - b(h_m + v)| &\leq |b(h_m + u) - c(u)| \\
&\quad + |c(u) - c(v)| + |c(v) - b(h_m + v)| \\
&\leq \varepsilon/2 + \mathrm{osc}_U(c) + \varepsilon/2,
\end{aligned}$$

and consequently, $\mathrm{osc}_{h_m+U}(b) \leq \mathrm{osc}_U(c) + \varepsilon$. Completely analogously, we derive $\mathrm{osc}_U(c) \leq \mathrm{osc}_{h_m+U}(b) + \varepsilon$, and taking this together we see that

$$|\mathrm{osc}_{h_m+U}(b) - \mathrm{osc}_U(c)| < \varepsilon \qquad \forall m > m_0$$

which proves our claim. □

Definition 3.31. *We call a function $b \in L^\infty$ rich if M_b is a rich operator. Otherwise we call b ordinary. There are even functions $b \in L^\infty$ for which no sequence $h \to \infty$ in $\mathbb{Z}^n$ at all leads to a limit operator of M_b. Such functions will be referred to as poor functions. We denote the set of rich functions by $L_\$^\infty$.*

Proposition 3.32. *$L_\$^\infty$ is a Banach subalgebra of L^∞.*

Proof. From Proposition 3.11 a) we get that

$$\{M_b : b \in L_\$^\infty\} = \{M_b : b \in L^\infty\} \cap L_\$(E, \mathcal{P})$$

is a Banach subalgebra of $L(E)$ and hence of $\{M_b : b \in L^\infty\}$, which implies our claim. □

Example 3.33. a) Let $n = 1$. The function

$$b(x) = (-1)^{[x/\pi]}, \qquad x \in \mathbb{R},$$

where $[y]$ denotes the integer part of $y \in \mathbb{R}$, is a poor function. b is 2π-periodic and has a jump at every multiple of π. Since π is irrational, the sequence $(b|_{h_m+U})_m$ cannot be a Cauchy sequence in $L^\infty(U)$ for any sequence of integers (h_m) if $U \supset 2Q$. The reason for b being poor is clearly the condition that all h_m in Definition 3.3 have to be integers.

b) Again let $n = 1$. Another example of a poor function is

$$b(x) = \sin(x^2), \qquad x \in \mathbb{R}.$$

One can easily show that no sequence (of integers or reals) tending to infinity leads to a limit operator of M_b. The reason is that $b \in BC$ is not uniformly continuous. We will show in Proposition 3.38 that functions in $BC \setminus BUC$ cannot be rich. □

3.4.2 Step Functions

Roughly speaking, step functions are piecewise constant functions on a lattice of hypercubes.

Definition 3.34. *A function $f \in L^\infty$ is called step function with steps of size $s > 0$ if there is an $x_0 \in \mathbb{R}^n$ such that f is constant on all cubes*

$$H_\alpha := x_0 + s(\alpha + H), \qquad \alpha \in \mathbb{Z}^n. \tag{3.19}$$

The set of these functions will be denoted by T_s. Finally, put

$$T_\mathbb{Q} := \bigcup_{p,q \in \mathbb{N}} T_{p/q}.$$

Functions in T_1 can be identified with elements of $\ell^\infty(\mathbb{Z}^n, \mathbb{C})$ in the natural way. The function b in Example 3.33 a) is obviously in T_π, and it has proven to be ordinary (even poor). We will see that this follows from the irrationality of π.

Proposition 3.35. *Step functions with rational step size are rich, $T_\mathbb{Q} \subset L^\infty_\$$.*

Proof. Pick some arbitrary $p, q \in \mathbb{N}$. Since $T_{p/q} \subset T_{1/q}$, it remains to show that all functions in $T_{1/q}$ are rich. So take $b \in T_s$ with $s = 1/q$, and let $x_0 \in \mathbb{R}^n$ be such that b is constant on every cube (3.19). The function b can easily be identified with a sequence $(b_\alpha) \in \ell^\infty(\mathbb{Z}^n, \mathbf{X})$ with $\mathbf{X} = \mathbb{C}^{q^n}$, where, for every $\alpha \in \mathbb{Z}^n$, the vector $b_\alpha \in \mathbb{C}^{q^n}$ contains the function values of b at the q^n steps inside the size 1 cube $x_0 + \alpha + H$.

By this simple construction, we can identify M_b with the generalized multiplication operator with symbol $(b_\alpha I_\mathbf{X})$ acting on $\ell^p(\mathbb{Z}^n, \mathbf{X})$. Since $\mathbf{X} = \mathbb{C}^{q^n}$ is finite-dimensional, our claim follows directly from Corollary 3.23. $\square$

3.4.3   Bounded and Uniformly Continuous Functions

Remember that BC and BUC denote the Banach algebras of all continuous and all uniformly continuous functions in L^∞, respectively.

Proposition 3.36. *Every function $b \in$ BUC is rich. Moreover, if M_c is a limit operator of M_b, then $c \in$ BUC as well.*

Proof. Take an arbitrary $b \in$ BUC and some $\varepsilon > 0$. Then there is a $\delta > 0$ such that

$$\mathrm{osc}_{x+\delta Q}(b) \leq \varepsilon \qquad \text{for all } x \in \mathbb{R}^n. \tag{3.20}$$

So this is true for every $\delta' \in \mathbb{Q}$ with $0 < \delta' \leq \delta$ as well. Consequently, there is a step function $f \in T_{\delta'}$ with $\|b - f\|_\infty < \varepsilon$ which tells that $T_\mathbb{Q}$ is dense in BUC. Propositions 3.35 and 3.32 show that BUC $\subset L^\infty_\$$.

Now suppose M_c is the limit operator of M_b with respect to some sequence $h = (h_m)$ tending to infinity. Take an arbitrary $\varepsilon > 0$, and choose $\delta > 0$ such that (3.20) holds. Then we have

$$\mathrm{osc}_{h_m+x+\delta Q}(b) \leq \varepsilon, \qquad x \in \mathbb{R}^n, \, m \in \mathbb{N},$$

and Lemma 3.30 shows that $\mathrm{osc}_{x+\delta Q}(c) \leq \varepsilon$ for all $x \in \mathbb{R}^n$, i.e. $c \in$ BUC. $\square$

So all uniformly continuous BC-functions are rich. But are there any more rich functions among the others in BC? To answer this question, we first have a look at an example of such a function in BC $\setminus$ BUC.

Example 3.37. Take a continuous function $f : \mathbb{R} \to [0, 1]$ which is only supported in the intervals $m^2 + (-\frac{1}{m}, \frac{1}{m})$ and is subject to $f(m^2) = 1$ for all $m \in \mathbb{N}$. M_f has no limit operator with respect to any subsequence of $h = (m^2)$. But there are sequences like $(m^2 + m)$ which lead to a limit operator of M_f. So f is not poor. However, f is just ordinary. $\square$

Indeed, one can show that every function in $BC \setminus BUC$ is a little bit like f and thus ordinary.

Proposition 3.38. *There are no rich functions in* $BC \setminus BUC$.

Proof. Take a bounded and – not uniformly – continuous function b. Since b is uniformly continuous on every compactum, the reason for b not being in BUC lies at infinity, i.e. there is an $\varepsilon_0 > 0$ and a sequence $(x_m) \subset \mathbb{R}^n$ with $x_m \to \infty$ and

$$\operatorname{osc}_{x_m + \frac{1}{m}Q}(b) \geq \varepsilon_0 \tag{3.21}$$

for all $m = 1, 2, \ldots$

Now choose the sequence $h = (h_m) := ([x_m]) \subset \mathbb{Z}^n$. We will see that M_b does not possess a limit operator with respect to any subsequence of h. Suppose M_c is the limit operator of the subsequence (h_{m_k}) of $h = (h_m)$. Then b is uniformly continuous on every compactum $h_{m_k} + 2Q$, $k \in \mathbb{N}$, and (3.18) shows that c is uniformly continuous on $2Q$. But then, there is a $\delta \in (0, 1)$ such that

$$\operatorname{osc}_{x + \delta Q}(c) \leq \frac{\varepsilon_0}{2}, \qquad x \in Q \tag{3.22}$$

since $x + \delta Q \subset 2Q$. For every $m > 1/\delta$ and $x := x_m - h_m \in Q$, inequalities (3.21) and (3.22) show that $b|_{h_m + 2Q}$ and $c|_{2Q}$ differ by at least $\varepsilon_0/4$ on a subset of $2Q$ with positive measure since

$$\operatorname{osc}_{h_m + x + \delta Q}(b) \geq \varepsilon_0 \qquad \text{and} \qquad \operatorname{osc}_{x + \delta Q}(c) \leq \frac{\varepsilon_0}{2}.$$

But this is a contradiction to $b|_{h_{m_k} + 2Q} \to c|_{2Q}$ as $k \to \infty$. $\square$

From Propositions 3.36 and 3.38 we conclude the following characterization of the class BUC.

Proposition 3.39. *A function* $f \in BC$ *is rich if and only if* $f \in BUC$. *In short,*

$$BUC = BC \cap L_{\$}^{\infty}.$$

3.4.4 Intermezzo: Essential Cluster Points at Infinity

In the previous two subsections we dealt with piecewise constant and with continuous functions. The evaluation of an arbitrary function $f \in L^{\infty}$ requests some more care.

Take a function $f \in L^{\infty}$ and a measurable set $U \subset \mathbb{R}^n$. How can we know whether or not a given number $c \in \mathbb{C}$ is *attained* by f in U? Since we do not distinguish between two functions in L^{∞} which differ in a set of measure zero only, it makes sense to accept only those values $c \in \mathbb{C}$ as actually *attained* by f in U, and write $c \in f(U)$, for which the set

$$\{x \in U \ : \ |f(x) - c| < \varepsilon\}$$

has a positive measure for every $\varepsilon > 0$. A more elegant way to say the same thing is that $f(U)$ is the spectrum of $f|_U$ in the Banach algebra $L^\infty(U)$. One refers to $f(U)$ as the *essential range of f in U*.

For the study of multiplication operators with symbol $f \in L^\infty$ and their limit operators, it is sufficient to evaluate f in a neighbourhood of infinity. Therefore, if $s \in S^{n-1}$, we denote by $\mathcal{U}$ the set of all neighbourhoods U_R^∞ of infinity (3.3) with $R > 0$, and by $\mathcal{U}_s$ we denote the set of all neighbourhoods $U_{R,V}^\infty$ of infinity (3.4) with $R > 0$ and a neighbourhood V of s in S^{n-1}. Then put

$$f(\infty) := \bigcap_{U \in \mathcal{U}} f(U) \quad \text{and} \quad f(\infty_s) := \bigcap_{U \in \mathcal{U}_s} f(U). \tag{3.23}$$

We call these sets the *local essential ranges* and their elements the *essential cluster points* of f at ∞ and ∞_s, respectively.

Lemma 3.40. *A number $c \in \mathbb{C}$ is contained in $f(\infty_s)$ for a given function $f \in L^\infty$ and a direction $s \in S^{n-1}$ if and only if there is a sequence $(x_m)_{m=1}^\infty \subset \mathbb{R}^n$ tending to ∞_s such that*

$$\mathrm{dist}\Big(c,\, f(x_m + Q) \Big) \to 0 \quad \text{as} \quad m \to \infty. \tag{3.24}$$

The same statement is true if we replace ∞_s by ∞.

Proof. Let $f \in L^\infty$ and fix a direction $s \in S^{n-1}$. The proof for ∞ is analogously.

First suppose there is a sequence $(x_m) \subset \mathbb{R}^n$ tending to ∞_s with (3.24). For every neighbourhood $U \in \mathcal{U}_s$, there is an $m_0 \in \mathbb{N}$ such that $x_m + Q \subset U$ for all $m \geq m_0$, and hence $W := \cup_{m \geq m_0} f(x_m + Q) \subset f(U)$. Since spectra are closed, the set $f(U)$ is closed, and from (3.24) we get that $c \in \mathrm{clos}\, W \subset f(U)$.

Now suppose $c \in f(\infty_s)$. For every $m \in \mathbb{N}$, put $V_m(s) = \{x \in S^{n-1} : |x - s|_2 < 1/m\}$. By (3.23), for every $m \in \mathbb{N}$, there exists a set $W_m \subset U_{m, V_m(s)}^\infty$ with $|W_m| > 0$ and $\|f|_{W_m} - c\|_\infty < 1/m$. If we choose $x_m \in W_m$ such that $|(x_m + Q) \cap W_m| > 0$, then $x_m \to \infty_s$ and (3.24) holds since $\mathrm{dist}\,(c, f(x_m + Q)) \leq \|f|_{W_m} - c\|_\infty < 1/m$. $\square$

Proposition 3.41. *For every $f \in L^\infty$, the identity*

$$f(\infty) = \bigcup_{s \in S^{n-1}} f(\infty_s)$$

holds.

Proof. Clearly, $f(\infty_s) \subset f(\infty)$ for every $s \in S^{n-1}$.

Conversely, if $c \in f(\infty)$, then, by Lemma 3.40, there exists a sequence $(x_m) \subset \mathbb{R}^n$ tending to ∞ such that (3.24) holds. But since S^{n-1} is compact, there is a subsequence (x_{m_k}) of (x_m) for which $x_{m_k}/|x_{m_k}|_2$ converges to some $s \in S^{n-1}$. But then, Lemma 3.40 again, now applied to the subsequence with $x_{m_k} \to \infty_s$, shows that $c \in f(\infty_s)$. $\square$

3.4.5 Slowly Oscillating Functions

Here we will study another interesting class of functions which have the property that limit operators of their multiplication operators have an especially simple structure.

Definition 3.42. *We say that a function $f \in L^\infty$ is slowly oscillating and write $f \in$ SO if $osc_x(f) \to 0$ as $x \to \infty$. Moreover, for every $s \in S^{n-1}$, we will say that $f \in L^\infty$ is slowly oscillating towards ∞_s and write $f \in SO_s$ if $osc_x(f) \to 0$ as $x \to \infty_s$.*

Similar to Propositions 3.9 and 3.41, we can prove the following result.

Proposition 3.43. *The identity*

$$\text{SO} = \bigcap_{s \in S^{n-1}} \text{SO}_s$$

holds.

Proof. If $f \in$ SO and $s \in S^{n-1}$, then clearly $osc_x(f) \to 0$ if $x \to \infty_s$, whence $f \in SO_s$.

Now suppose $f \notin$ SO. Then there exists a sequence $(x_m) \subset \mathbb{R}^n$ with $x_m \to \infty$ and an $\varepsilon_0 > 0$ with

$$|osc_{x_m}(f)| \geq \varepsilon_0 \tag{3.25}$$

for all $m \in \mathbb{N}$. Since S^{n-1} is compact, we can choose a subsequence (x_{m_k}) from (x_m) which tends to ∞_s for some $s \in S^{n-1}$. But then (3.25) shows that $f \notin SO_s$. $\square$

Proposition 3.44. SO *and* SO_s *are Banach subalgebras of L^∞ for every $s \in S^{n-1}$.*

Proof. This proof is very straightforward. Just note that

$$
\begin{aligned}
osc_x(f + g) &\leq osc_x(f) + osc_x(g), \\
osc_x(fg) &\leq osc_x(f)\,\|g\| + \|f\|\,osc_x(g) \quad \text{and} \\
osc_x(f) &\leq osc_x(f_m) + 2\,\|f - f_m\|
\end{aligned}
$$

hold for all $f, g, f_m \in L^\infty$ and $x \in \mathbb{R}^n$. $\square$

Lemma 3.45. *Let f be an arbitrary function in L^∞.*

a) *For every $s \in S^{n-1}$, the following three conditions are equivalent.*

   (i) $f \in SO_s$,

   (ii) $\lim_{x \to \infty_s} osc_x(f) = 0$,

   (iii) $\lim_{x \to \infty_s} osc_{x+U}(f) = 0$ *for all bounded and measurable $U \subset \mathbb{R}^n$.*

b) *If f is differentiable in some neighbourhood of ∞_s, then the convergence of its gradient $\nabla f(x) \to 0$ as $x \to \infty_s$ is sufficient for $f \in SO_s$ – but not necessary.*

Proof. a) (i) $\Longleftrightarrow$ (ii) holds by Definition 3.42.
(iii)$\Rightarrow$(ii) is trivial since (ii) is just

$$\lim_{x \to \infty_s} osc_{x+U}(f) = 0 \qquad\qquad (3.26)$$

with $U = Q$.

(ii)$\Rightarrow$(iii): If (ii) holds, then we have (3.26) for $U = Q$ and all subsets of Q. If (3.26) holds for a set U, then it also holds for all sets of the form $t + U$, $t \in \mathbb{R}^n$ in place of U. Finally, if it holds for $U = U_1$ and $U = U_2$ with $|U_1 \cap U_2| > 0$, then it clearly holds for $U = U_1 \cup U_2$ since $osc_{x+(U_1 \cup U_2)}(f) \le osc_{x+U_1}(f) + osc_{x+U_2}(f)$. Taking all this together, it is clear that (3.26) holds for all bounded and measurable $U \subset \mathbb{R}^n$.

b) Take an $\varepsilon > 0$. If $\nabla f(x) \to 0$ as $x \to \infty_s$, there is some neighbourhood V_ε of ∞_s such that $|\nabla f(x)| < \varepsilon/(2n)$ for all $x \in V_\varepsilon$. But from

$$|f(x+u) - f(x+v)| = |\nabla f(\xi_{x,u,v}) \cdot (u - v)| < \varepsilon, \qquad u, v \in Q, \quad \xi_{x,u,v} \in x + Q,$$

we conclude that $osc_x(f) \le \varepsilon$ if $x \in V_\varepsilon$. This is (ii).

Check the function f in Example 3.46 d) to see that $\nabla f(x)$ need not tend to zero as $x \to \infty$ if $f \in SO$. $\square$

In what follows, we will often use property (iii) of Lemma 3.45 to characterize the sets SO_s.

Example 3.46. a) The simplest examples of slowly oscillating functions are functions which converge at ∞. If $f \in L^\infty$ only converges towards ∞_s for some $s \in S^{n-1}$, then at least $f \in SO_s$.

b) Take $n = 2$ and

$$f(x, y) = \frac{\sin x}{y^2 + 1}, \qquad x, y \in \mathbb{R}.$$

Then $f(x)$ converges as $x \to \infty_s$, and hence $f \in SO_s$, if $s \in S^1 \setminus \{(-1, 0), (1, 0)\}$. But $f \notin SO_s$ for $s \in \{(-1, 0), (1, 0)\}$.

c) Take $n = 1$ and $f(x) = \sin \sqrt{|x|}$. Then $f \in SO$ since $f'(x) \to 0$ as $x \to \infty$.

d) Take $n = 1$ and

$$f(x) = \frac{\sin x^2}{x}, \qquad x \in \mathbb{R}.$$

Then $f(x)$ converges as $x \to \infty$, whence $f \in SO$. But note that $f'(x)$ does not tend to zero as $x \to \infty$ although $f \in SO$. Slowly oscillating functions need not even be continuous. For instance, look at

$$g(x) = \frac{(-1)^{[x]}}{x}, \qquad x \in \mathbb{R},$$

which converges for $x \to \infty$ but has a jump at every $x \in \mathbb{Z}$. $\qquad\square$

We will now study the set of limit operators of M_b when b is slowly oscillating.

Proposition 3.47. *If $s \in S^{n-1}$ and $b \in SO_s$, then the local operator spectrum of M_b is*

$$\sigma_s^{\mathrm{op}}(M_b) = \{\, cI \; : \; c \in b(\infty_s) \,\}.$$

Proof. Pick some $c \in b(\infty_s)$, and take an $\varepsilon > 0$. By Lemma 3.40, there is a sequence $(x_m) \subset \mathbb{R}^n$ tending to ∞_s and a sequence $(c_m) \subset \mathbb{C}$ with $c_m \in b(x_m+Q)$ and $c_m \to c$ as $m \to \infty$. Since $b \in SO_s$, there is an $m_0 \in \mathbb{N}$ such that, for every $m > m_0$, we have $\mathrm{osc}_{x_m+2Q}(b) < \varepsilon/2$. If m_0 is taken large enough that, in addition, $|c_m - c| < \varepsilon/2$ holds, then

$$|b(x_m + x) - c| \; \le \; |b(x_m + x) - c_m| + |c_m - c| \; \le \; \mathrm{osc}_{x_m+2Q}(b) + |c_m - c| \; < \; \varepsilon$$

for almost all $x \in 2Q$, i.e. $\|b|_{x_m+2Q} - c\|_\infty < \varepsilon$ if $m > m_0$. Now define the sequence $h = (h_m) \subset \mathbb{Z}^n$ by $h_m := [x_m]$. Then $h_m + Q$ is contained in $x_m + 2Q$, and consequently, $\|b|_{h_m+Q} - c\|_\infty < \varepsilon$ for all $m > m_0$. From Proposition 3.27 we get that cI is the limit operator of M_b with respect to the sequence h.

Conversely, by Lemma 3.30, it is clear that, for all limit operators M_c of M_b towards ∞_s, the local oscillation $\mathrm{osc}_x(c)$ has to be zero at every $x \in \mathbb{R}^n$, i.e. c is a constant. From Proposition 3.27 and Lemma 3.40 we get that $c \in b(\infty_s)$. $\qquad\square$

Proposition 3.9 and 3.41 allow us to glue these local results together and get the following[3].

Corollary 3.48. *If $b \in SO$, then $\sigma^{\mathrm{op}}(M_b) = \{cI : c \in b(\infty)\}$.*

Proof. If $b \in SO$, then $b \in SO_s$ for all $s \in S^{n-1}$ by Proposition 3.43. By Propositions 3.9, 3.47 and 3.41, we have

$$\sigma^{\mathrm{op}}(M_b) = \bigcup_{s \in S^{n-1}} \sigma_s^{\mathrm{op}}(M_b) = \bigcup_{s \in S^{n-1}} \{cI : c \in b(\infty_s)\} = \{cI : c \in b(\infty)\}. \qquad\square$$

Proposition 3.49. *Slowly oscillating functions are rich, $SO \subset L_\$^\infty$.*

Proof. Take a function $b \in SO$ and an arbitrary sequence $h = (h_m) \subset \mathbb{Z}^n$ with $h_m \to \infty$. Now pick a sequence (c_m) of complex numbers such that $c_m \in b(h_m+Q)$ for every $m \in \mathbb{N}$. Since $(c_m) \subset \mathbb{C}$ is bounded, there is a subsequence g of h such that the corresponding subsequence of (c_m) converges to some $c \in \mathbb{C}$. Now proceed as in the proof of Proposition 3.47, showing that cI is the limit operator of M_b with respect to g. $\qquad\square$

The following proposition shows that, in a sense, even the reverse of Corollary 3.48 is true.

---

[3]Of course, Corollary 3.48 also can be shown directly, exactly like Proposition 3.47.

Proposition 3.50. *If $b \in L_{\$}^{\infty}$ and every limit operator of M_b is a multiple of the identity operator I, then $b \in$ SO.*

Proof. Let the conditions of the proposition be fulfilled, and suppose that $b \notin$ SO. Then there exist an $\varepsilon_0 > 0$ and a sequence of points $(x_m) \subset \mathbb{R}^n$ tending to infinity such that, for every $m \in \mathbb{N}$, we have $\mathrm{osc}_{x_m + Q}(b) \geq \varepsilon_0$.

Now put $h = (h_m) \subset \mathbb{Z}^n$ with $h_m := [x_m]$. Since $x_m + Q \subset h_m + 2Q$, we have

$$\mathrm{osc}_{h_m + 2Q}(b) \geq \mathrm{osc}_{x_m + Q}(b) > \varepsilon_0$$

for all $m \in \mathbb{N}$. Since $b \in L_{\$}^{\infty}$, there is a subsequence g of h such that the limit operator of M_b exists with respect to g. Denote this limit operator by M_c. By Lemma 3.30, we then conclude that $\mathrm{osc}_{2Q}(c) \geq \varepsilon_0$, and hence, c is certainly not constant on $2Q$. So at least one limit operator of M_b is not a multiple of I which contradicts our assumption. $\qquad\square$

These results motivate the following definition.

Definition 3.51. *We denote the set of functions $b \in L^{\infty}$, for which every limit operator of M_b is a multiple of the identity, by CL; that is*

$$\mathrm{CL} := \left\{ b \in L^{\infty} : \sigma^{\mathrm{op}}(M_b) \subset \{cI : c \in \mathbb{C}\} \right\}.$$

Now we can summarize Corollary 3.48 and Propositions 3.49, 3.50 by the following result.

Proposition 3.52. *A function $b \in L^{\infty}$ is slowly oscillating if and only if it is rich and every limit operator of M_b is a multiple of the identity. In short,*

$$\mathrm{SO} = \mathrm{CL} \cap L_{\$}^{\infty}.$$

Remark 3.53. The notion of a slowly oscillating function is easily generalized to elements of $\ell^{\infty}(\mathbb{Z}^n, L(\mathbf{X}))$ with an arbitrary Banach space $\mathbf{X}$. We say that $b = (b_\alpha)_{\alpha \in \mathbb{Z}^n}$ is *slowly oscillating* if

$$\|b_{\alpha + q} - b_\alpha\|_{L(\mathbf{X})} \to 0 \qquad \text{as} \qquad \alpha \to \infty$$

for all $q \in Q \cap \mathbb{Z}^n$ or, equivalently, for all $q \in \mathbb{Z}^n$. Analogously, one can study slow oscillation towards ∞_s for $s \in S^{n-1}$. An obviously equivalent description of this fact is that, for all $q \in \mathbb{Z}^n$, the sequence $V_q b - b$ vanishes at ∞ or ∞_s, respectively.

It is readily checked that, in the case $\mathbf{X} = \mathbb{C}$, all results of this subsection carry over to generalized multiplication operators with a slowly oscillating symbol. As a consequence, in this case we get that, for $A \in \mathrm{BDO}_{\$}^p$, every limit operator of A is shift-invariant if and only if A has slowly oscillating coefficients!

Note that, if $\mathbf{X}$ is infinite-dimensional, then $\hat{M}_b$ no longer needs to be rich if b is slowly oscillating. But however, even then it holds that all limit operators of $\hat{M}_b$ are generalized multiplication operators with a constant symbol. $\qquad\square$

3.4.6 Admissible Additive Perturbations

We know that neither the operator spectrum of M_b nor the property whether or not M_b is rich is affected by additive perturbations $T \in K(E, \mathcal{P})$. Since our focus is on multiplication operators, we will have a look at the set of all functions $f \in L^\infty$ for which $T = M_f \in K(E, \mathcal{P})$. We denote this set by

$$
\begin{aligned}
L_0^\infty &:= \{f \in L^\infty : M_f \in K(E, \mathcal{P})\} \\
&= \{f \in L^\infty : Q_m f \to 0 \text{ as } m \to \infty\}.
\end{aligned}
$$

Equivalently, L_0^∞ is the set of all functions $f \in L^\infty$ with local essential range $f(\infty) = \{0\}$; that is $\lim_{x \to \infty} f(x) = 0$. Also note that $L_0^\infty = (L^\infty)_0$ in the notation of (1.13). Moreover, it is well-known that $(L_0^\infty)^* = L^1$, and hence $(L_0^\infty)^{**} = L^\infty$.

There is another characterization of L_0^∞ in terms of limit operators.

Proposition 3.54. $f \in L_0^\infty$ *if and only if* f *is rich and* $\sigma^{\mathrm{op}}(M_f) = \{0\}$.

Proof. If $f \in L_0^\infty$, then clearly, f is rich with all limit operators of M_f being 0. The reverse implication follows from Proposition 3.50 and Corollary 3.48. $\quad\square$

It is immediate that L_0^∞ is a closed ideal in L^∞, and that the limit operators of M_b and its property of being rich only depend on the coset $b + L_0^\infty$ in L^∞/L_0^∞.

Definition 3.55. *If F is a subset of L^∞, then we refer to $F + L_0^\infty$ as the* relaxation *of F.*

Suppose, for a set $F \subset L^\infty$, we know something about the limit operators of M_b for all $b \in F$. Then we still have the same knowledge (and the same limit operators) if we relax the condition $b \in F$ to $b \in F + L_0^\infty$, which can enlarge the class of functions quite considerably – think of $F = \mathrm{BUC}$, for example.

3.4.7 Slowly Oscillating and Continuous Functions

Sometimes, the class

$$ \mathrm{SOC} := \mathrm{SO} \cap \mathrm{BC} $$

of slowly oscillating and continuous functions is of interest.

Proposition 3.56. $\mathrm{SOC} \subset \mathrm{BUC}$

Proof. Although there is a direct proof (using some ε's and δ's), we will do something different: By Proposition 3.49, we have $\mathrm{SO} \subset L_\$^\infty$. Consequently,

$$ \mathrm{SOC} = \mathrm{SO} \cap \mathrm{BC} \subset L_\$^\infty \cap \mathrm{BC} = \mathrm{BUC}, $$

by Proposition 3.39. $\quad\square$

By definition, SOC comes from SO by taking intersection with BC. Conversely, SO can be derived from SOC by adding L_0^∞.

Proposition 3.57. SOC $+ L_0^\infty =$ SO

Proof. If $f \in$ SOC and $g \in L_0^\infty$, then both are in SO, and hence, $f + g \in$ SO.

For the reverse inclusion, take an arbitrary $f \in$ SO. Now define a function g as follows: In the integer points $x \in \mathbb{Z}^n$, let $g(x)$ be some value from the essential range $f(x + H)$, and then use an interpolation idea by setting $g(x + h)$ a convex combination of the function values of g in the 2^n corners of the hypercube $x + H$:

$$g(x + h) := \sum_{v \in \{0,1\}^n} \left(\prod_{i=1}^n (1 - |h_i - v_i|) \right) g(x + v),$$

where $h = (h_i) \in H$ and $v = (v_i) \in \{0, 1\}^n$. Then g is continuous, and

$$g(x + H) \subset \text{conv}\, f(x + 2H) \qquad \text{for all} \qquad x \in \mathbb{Z}^n,$$

by our construction. Consequently, $\text{osc}_{x+H}(g) \le \text{osc}_{x+2H}(f) \to 0$ as $x \to \infty$, whence $g \in$ SOC. Finally, from $\|(f - g)|_{x+H}\|_\infty \le \text{osc}_{x+2H}(f) \to 0$ as $x \to \infty$, we get $f - g \in L_0^\infty$, whence $f \in$ SOC $+ L_0^\infty$. $\qquad\qquad\square$

Proposition 3.57 shows that SO is the relaxation of SOC. From Example 3.46 a) we know that the following class is an especially simple subset of SOC. By $C(\overline{\mathbb{R}^n})$ we denote the set of all continuous functions $f : \mathbb{R}^n \to \mathbb{C}$ for which the limits

$$\lim_{x \to \infty_s} f(x)$$

exist for all $s \in S^{n-1}$, and by $C(\dot{\mathbb{R}}^n)$ we denote its subset of functions f for which this limit does not depend on s. Clearly,

$$C(\dot{\mathbb{R}}^n) \subset C(\overline{\mathbb{R}^n}) \subset \text{SOC} \subset \text{BUC}$$

are all Banach subalgebras of L^∞.

We will come back to SO and SOC later in Subsection 3.4.11.

3.4.8 Almost Periodic Functions

This time we will start in the discrete case, and we pass to the continuous case towards the end of the subsection. Suppose $\mathbf{X}$ and $\mathbf{Y}$ are arbitrary Banach spaces. For a generalized multiplication operator $\hat{M}_b$ on $\ell^p(\mathbb{Z}^n, \mathbf{X})$ with $b \in \ell^\infty(\mathbb{Z}^n, L(\mathbf{X}))$, it clearly holds that

$$V_{-\gamma} \hat{M}_b V_\gamma = \hat{M}_{V_{-\gamma} b}, \qquad \gamma \in \mathbb{Z}^n. \tag{3.27}$$

We say that $b \in \ell^\infty(\mathbb{Z}^n, \mathbf{Y})$ is *periodic* if there are linearly independent vectors $\omega_1, \dots, \omega_n \in \mathbb{Z}^n$ such that

$$V_{\omega_k} b = b, \qquad k = 1, \dots, n.$$

If we put $M := [\omega_1, \ldots, \omega_n] \in \mathbb{Z}^{n \times n}$, then $S := M(H) \cap \mathbb{Z}^n$ is the discrete parallelotope spanned by $\omega_1, \ldots, \omega_n$, and b is completely determined by the finitely many values b_α with $\alpha \in S$. Consequently, the set

$$\{V_\gamma b\}_{\gamma \in \mathbb{Z}^n} = \{V_\gamma b\}_{\gamma \in S}$$

is finite, and, by (3.27), so is the set $\{V_{-\gamma} \hat{M}_b V_\gamma\}_{\gamma \in \mathbb{Z}^n}$. Both sets have (at most) $\#S = |\det M|$ elements.

A slightly weaker property than periodicity is the following.

Definition 3.58. *An element $b \in \ell^\infty(\mathbb{Z}^n, \mathbf{Y})$ is called almost periodic if the set $\{V_\gamma b\}_{\gamma \in \mathbb{Z}^n}$ is relatively compact in $\ell^\infty(\mathbb{Z}^n, \mathbf{Y})$.*

This relative compactness is still a very strong property. It ensures that, if $A = \hat{M}_b$ with an almost periodic symbol b, and $h = (h_m) \subset \mathbb{Z}^n$ tends to infinity, then there is a subsequence g of h for which the sequence $V_{-g_m} A V_{g_m}$ converges as $m \to \infty$ – even in the operator norm! So if b is almost periodic, then $A = \hat{M}_b$ is rich, and the $\mathcal{P}$-convergence of the sequence (3.5) is actually norm-convergence.

This fact carries over to BDO^p with almost periodic coefficients.

Corollary 3.59. *If A is a band-dominated operator with almost periodic coefficients, then A is rich, and the invertibility of any limit operator A_h of A already implies the invertibility of A.*

Proof. That A is rich follows from the above and Proposition 3.11. If $h = (h_m)$ goes to infinity and A_h is invertible, then, by Lemma 1.2 b) and by what was said above, $V_{-h_m} A V_{h_m}$ is invertible for sufficiently large $m \in \mathbb{N}$, and hence A itself is invertible. $\qquad\square$

For $\varepsilon > 0$, we say that $\omega \in \mathbb{Z}^n$ is an ε-almost period of $b \in \ell^\infty(\mathbb{Z}^n, \mathbf{Y})$, if

$$\|V_\omega b - b\|_\infty < \varepsilon,$$

and we denote the set of all ε-almost periods of b by $\Omega_\varepsilon(b)$.

Proposition 3.60. *$b \in \ell^\infty(\mathbb{Z}^n, \mathbf{Y})$ is almost periodic if and only if, for every $\varepsilon > 0$, the set $\Omega_\varepsilon(b)$ is relatively dense in $\mathbb{Z}^n$; that is, there exists a bounded set $S_\varepsilon \subset \mathbb{Z}^n$ such that every translate of S_ε in $\mathbb{Z}^n$ contains an ε-almost period of b.*

Proof. The proof for $n = 1$ can be found in [13] (Theorem 1.26) for example. The result is easily carried over to arbitrary $n \in \mathbb{N}$. $\qquad\square$

As a consequence of Proposition 3.60, we get the following result.

Proposition 3.61. *If $b \in \ell^\infty(\mathbb{Z}^n, L(\mathbf{X}))$ is almost periodic, then*

$$\sigma_s^{\mathrm{op}}(\hat{M}_b) = \sigma^{\mathrm{op}}(\hat{M}_b)$$

for all $s \in S^{n-1}$.

Proof. Let $b \in \ell^\infty(\mathbb{Z}^n, L(\mathbf{X}))$ be almost periodic, take arbitrary $s, t \in S^{n-1}$, and suppose $A \in \sigma_s^{\mathrm{op}}(\hat{M}_b)$. We will show that $A \in \sigma_t^{\mathrm{op}}(\hat{M}_b)$ as well, which, by Proposition 3.9, proofs this proposition.

From $A \in \sigma_s^{\mathrm{op}}(\hat{M}_b)$ and Proposition 3.6 d) we get that $A = \hat{M}_c$ and that there exists a sequence $h = (h_m) \subset \mathbb{Z}^n$ with $h_m \to \infty_s$ and $V_{-h_m} b \overset{\mathcal{P}}{\to} c$ as $m \to \infty$ by (3.27). Since b is almost periodic, we can pass to a subsequence of h for which this $\mathcal{P}$-convergence $\overset{\mathcal{P}}{\to}$ is even convergence in L^∞. Suppose h is already this subsequence. Now choose a sequence $g = (g_m) \in \mathbb{Z}^n$ such that $g_m - h_m \in \Omega_{1/m}(b)$ and $g_m \to \infty_t$. Then, by Proposition 3.60,

$$
\begin{aligned}
\|V_{-g_m} b - c\|_\infty &\le \|V_{-g_m} b - V_{-h_m} b\|_\infty + \|V_{-h_m} b - c\|_\infty \\
&= \|V_{-g_m}(b - V_{g_m - h_m} b)\|_\infty + \|V_{-h_m} b - c\|_\infty \\
&= \|b - V_{g_m - h_m} b\|_\infty + \|V_{-h_m} b - c\|_\infty \\
&< 1/m + \|V_{-h_m} b - c\|_\infty \to 0
\end{aligned}
$$

as $m \to \infty$, and consequently $A = \hat{M}_c$ is the limit operator of $\hat{M}_b$ with respect to $g = (g_m)$ as well. $\qquad\square$

Now we come to the continuous case. If we identify a multiplication operator on L^p with its discretization (1.3); that is a generalized multiplication operator on $\ell^p(\mathbb{Z}^n, \mathbf{X})$ with $\mathbf{X} = L^p(H)$, we can carry over the notion of almost periodicity to L^∞.

Definition 3.62. *We say that $b \in L^\infty$ is $\mathbb{Z}$-almost periodic if the set $\{V_\gamma b\}_{\gamma \in \mathbb{Z}^n}$ is relatively compact in L^∞, and it is almost periodic if the set $\{V_\gamma b\}_{\gamma \in \mathbb{R}^n}$ is relatively compact in L^∞. The sets of all $\mathbb{Z}$-almost and almost periodic functions in L^∞ will be denoted by* $\mathrm{AP}_\mathbb{Z}$ *and* AP, *respectively.*

By what we discussed earlier, it follows that

$$
\mathrm{AP} \subset \mathrm{AP}_\mathbb{Z} \subset L_\$^\infty.
$$

The following example shows that AP is in fact a proper subset of $\mathrm{AP}_\mathbb{Z}$.

Example 3.63. If $n = 1$ and $s \in \mathbb{R}_+$, then

$$
f(x) = (-1)^{[x/s]}, \qquad x \in \mathbb{R}
$$

is a step function with step size s. It is easily seen that $f \in \mathrm{AP}_\mathbb{Z}$ if and only if $s \in \mathbb{Q}$. Using Proposition 3.35 and a little thought, it also follows that f is rich if and only if $s \in \mathbb{Q}$.

If we allow real-valued shift distances γ in $\{V_\gamma f\}$, then, for example, the sequence $(V_{k\sqrt{2}s} f)_{k \in \mathbb{N}}$ does not contain a convergent subsequence, regardless of the choice of s. So the set $\{V_\gamma f\}_{\gamma \in \mathbb{R}^n}$ is not relatively compact, whence $f \notin \mathrm{AP}$. $\qquad\square$

There is the following nice relationship between almost periodic functions in the discrete and the continuous case.

Lemma 3.64. *A sequence $b = (b_\alpha) \in \ell^\infty$ is almost periodic if and only if there is a function $f \in \mathrm{AP}$ with $b_\alpha = f(\alpha)$ for all $\alpha \in \mathbb{Z}^n$.*

Proof. For $n = 1$, this is shown in Theorem 1.27 of [13]. But this construction easily generalizes to $n \geq 1$. □

We prepare an equivalent characterization for functions in AP. Therefore, for every $a \in \mathbb{R}^n$, put

$$f_a(x) := \exp(i \langle a, x \rangle), \qquad x \in \mathbb{R}^n$$

where $\langle \cdot, \cdot \rangle$ refers to the standard inner product in $\mathbb{R}^n$. We denote by Π the set of all complex linear combinations of functions f_a with $a \in \mathbb{R}^n$, and we refer to elements of Π as trigonometric polynomials. Then the following holds.

Proposition 3.65. *For $f \in L^\infty = L^\infty(\mathbb{R}^n)$, the following conditions are equivalent.*

(i) *The set $\{V_\tau f\}_{\tau \in \mathbb{R}^n}$ is relatively compact in L^∞.*

(ii) *f is continuous, and, for every $\varepsilon > 0$, there is a cube $S \subset \mathbb{R}^n$ such that every translate of S contains a $\tau \in \mathbb{R}^n$ with $\|V_\tau f - f\|_\infty < \varepsilon$.*

(iii) *There is a sequence in Π which converges to f in the norm of L^∞.*

Proof. For $n = 1$ this is a well-known result by BOHR and BOCHNER which can be found in any textbook on almost periodic functions, for example [13] or [39, Chapter VI]. In fact, the equivalence of (i) and (iii) holds for locally compact abelian groups in place of $\mathbb{R}^n$ (see [39], page 192). The equivalence of (i) and (ii) is written down for $n = 1$ in Chapter VI, Theorem 5.5 of [39] which literally applies to $n \geq 1$ as well. □

As in the discrete case, the vectors τ in (ii) are called the ε-almost periods of f. From (iii) we get that AP is the smallest Banach subalgebra of L^∞ that contains all functions f_a with $a \in \mathbb{R}^n$,

$$\mathrm{AP} = \mathrm{clos}_{L^\infty} \Pi = \mathrm{closalg}_{L^\infty} \{f_a : a \in \mathbb{R}^n\}.$$

From this fact and $f_a \in \mathrm{BUC}$ for all $a \in \mathbb{R}^n$, we get that

$$\mathrm{AP} \subset \mathrm{BUC}. \tag{3.28}$$

Finally, refer to the Banach subalgebra of L^∞ that is generated by AP and $C(\overline{\mathbb{R}^n})$ as

$$\mathrm{SAP} := \mathrm{closalg}_{L^\infty} \{\mathrm{AP}, C(\overline{\mathbb{R}^n})\}.$$

The elements of SAP are called *semi-almost periodic functions*.

For $n = 1$, if $b_+ \in C(\overline{\mathbb{R}})$ with $b_+(+\infty) = 1$ and $b_+(-\infty) = 0$, and $b_- := 1 - b_+$ are fixed, then a famous result by SARASON [78] says that every function $f \in \mathrm{SAP}$ has a unique representation of the form

$$f = b_- f_- + b_0 + b_+ f_+ \tag{3.29}$$

with $f_+, f_- \in \mathrm{AP}$ and $b_0 \in C(\overline{\mathbb{R}})$ with $b_0(\pm\infty) = 0$.

In contrast to almost periodic functions, semi-almost periodic functions can have different almost periodic behaviour towards different directions of infinity. For dozens of very beautiful pictures of (semi-)almost periodic functions, see [7].

From (3.28) and $C(\overline{\mathbb{R}^n}) \subset \mathrm{BUC}$, we get that

$$\mathrm{SAP} \subset \mathrm{BUC} \subset L_{\$}^{\infty}.$$

From (3.29) and Proposition 3.61, we see that, for $n = 1$ and every $f \in \mathrm{SAP}$,

$$\sigma_-^{\mathrm{op}}(M_f) = \sigma_-^{\mathrm{op}}(M_{f_-}) = \sigma^{\mathrm{op}}(M_{f_-}) \quad \text{and} \quad \sigma_+^{\mathrm{op}}(M_f) = \sigma_+^{\mathrm{op}}(M_{f_+}) = \sigma^{\mathrm{op}}(M_{f_+})$$

with $f_+, f_- \in \mathrm{AP}$ uniquely determined by (3.29).

3.4.9 Oscillating Functions

In this subsection, we restrict ourselves to functions $b \in L^p$ on the axis; that is $n = 1$. Remember that $\mathbb{T}$ denotes the complex unit circle, and suppose $f : \mathbb{T} \to \mathbb{C}$ is a bounded and non-constant function. Then, for every $\omega > 0$, by

$$b(x) := f\left(\exp\left(2\pi i \frac{x}{\omega}\right)\right), \qquad x \in \mathbb{R}$$

we get a periodic function in L^∞ with $b(x + \omega) = b(x)$ for all $x \in \mathbb{R}$. In this case we will say that b oscillates with a constant frequency.

Furthermore, we will study functions whose frequency increases or decreases as x goes to infinity. Therefore, let $g : \mathbb{R} \to \mathbb{R}$ be an unbounded, odd and strictly monotonously increasing, differentiable function. Then put

$$b(x) := f\Big(\exp\big(2\pi i\, g(x)\big)\Big), \qquad x \in \mathbb{R}. \tag{3.30}$$

The three cases under consideration are

① $g'(x) \to +\infty$ as $x \to \infty$ (frequency tends to infinity),
② $g'(x) = 1/\omega$ for all $x \in \mathbb{R}$ (constant frequency – the periodic case),
③ $g'(x) \to 0$ as $x \to \infty$ (frequency tends to zero).

Proposition 3.66. *If f is continuous on $\mathbb{T}$, g is subject to one of the cases ①, ②, ③, and b is as in (3.30), then*

- *in case ①, b is always poor;*

- *in case ②, b is always rich with $\sigma^{\mathrm{op}}(M_b) = \{M_{V_c b} : c \in C\}$, where*

$$C = \begin{cases} \{0, \frac{1}{m}, \dots, \frac{k-1}{m}\} & \text{if } p = \frac{k}{m} \in \mathbb{Q} \text{ with } \gcd(k, m) = 1, \\ [0, p) & \text{if } p \in \mathbb{R} \setminus \mathbb{Q}; \end{cases} \tag{3.31}$$

- *and in case ③, b is always rich with $\sigma^{\mathrm{op}}(M_b) = \{cI : c \in f(\mathbb{T})\}$.*

Proof. The proof is a bit lengthy but rather straightforward. Firstly, it is shown that b is in $BC \setminus BUC$, AP and SO in case ①, ②, ③, respectively, and secondly, the operator spectra are derived from previous results on these function classes plus some additional arguments. The interested reader is referred to [49, Proposition 3.18] or [50, Proposition 2.25]. □

So if f is continuous, all answers in cases ①, ② and ③ are given – including an explicit description of the operator spectra. The situation changes a bit as soon as f has a single discontinuity, say a jump at $1 \in \mathbb{T}$:

- Case ① remains poor.

- Case ② is rich if and only if the period ω is rational.

- Most interesting is case ③. Here one cannot give a statement independently from the knowledge of the function g. This is demonstrated in Example 3.67.

Example 3.67. Consider the function

$$g(x) = \begin{cases} \log_a x, & x \ge e, \\ \frac{x}{e \ln a}, & -e < x < e, \\ \log_a(-x), & x \le -e, \end{cases}$$

where $a \ge e$ is fixed. It is easily seen that g is unbounded, odd, strictly monotonously increasing and differentiable, and $g'(x) \to 0$ as $x \to \infty$.

Then the function b from (3.30) jumps at $x = 0$ as well as at $x = \pm a^k$ for $k \in \mathbb{N}$, and it is continuous at every other point. The local oscillation of b, apart from near the jumps, goes to zero as x goes to infinity.

- Suppose the basis a is an integer. Then all jumps of b are at integer points. Now it is almost immediate that there is a step function b_s with step length 1 and a function $b_0 \in L_0^\infty$ such that $b = b_s + b_0$. As the sum of two rich functions, b is rich.

- Suppose $a = \sqrt{2}$. Then the jumps $x = \pm a^k$ of b are integers for k even and multiples of $\sqrt{2}$ for k odd. But then it is easily seen that the sequence $h = ([a^{2m+1}])_{m \in \mathbb{N}}$ has no subsequence leading to a limit operator of M_b, while the sequence $h = (a^{2m})_{m \in \mathbb{N}} = (2^m)_{m \in \mathbb{N}}$ leads to a limit operator. So here, b is ordinary – but not poor.

- If a is a transcendent number, then the jumps of b simply don't fit together modulo 1 because otherwise, we had integers k_1 and k_2 such that $a^{k_1} - a^{k_2}$ is an integer c, i.e. a solves the equation $x^{k_1} - x^{k_2} - c = 0$. So no subsequence of $h = ([a^m])_{m \in \mathbb{N}}$ leads to a limit operator, and hence, b is just ordinary.

For completeness, and this is independent from the explicit structure of g, we remark that b is never poor in case ③ since there are always sequences leading to a limit operator, for example, $h = ([g^{-1}(t+m)])_{m \in \mathbb{N}}$, where $g \circ g^{-1} = \mathrm{id} = g^{-1} \circ g$ and $t \in (0, 1)$ is fixed. □

It is not hard to see that the results from f having one jump can be extended to f having finitely many jumps on $\mathbb{T}$.

3.4.10 Pseudo-ergodic Functions

If we referred to slowly oscillating functions as those functions with especially simple associated limit operators, then the functions we consider now, in some sense show the opposite behaviour.

Fix a Banach space $\mathbf{Y}$ and a bounded and closed subset D of $\mathbf{Y}$. In accordance with DAVIES [26], we say that $b = (b_\alpha) \in \ell^\infty(\mathbb{Z}^n, \mathbf{Y})$ is *pseudo-ergodic with respect to D* if, for every finite set $S \subset \mathbb{Z}^n$, every function $c : S \to D$ and every $\varepsilon > 0$, there is a vector $\gamma \in \mathbb{Z}^n$ such that

$$\sup_{\alpha \in S} \|b_{\gamma + \alpha} - c_\alpha\|_{\mathbf{Y}} < \varepsilon. \qquad (3.32)$$

Large classes of random potentials have this property almost surely. A little thought reveals that there are in fact infinitely many vectors $\gamma \in \mathbb{Z}^n$ with (3.32) since this property also holds for all continuations of c to sets $S' \subset \mathbb{Z}^n$ containing S. Also it follows immediately from this definition that, if b is pseudo-ergodic with respect to D, then b is pseudo-ergodic with respect to every closed subset of D.

Roughly speaking, (b_α) is pseudo-ergodic with respect to D if one can find, up to a given precision $\varepsilon > 0$, any finite pattern of elements from D somewhere in (b_α). For example, it is conjectured[4] that the decimal expansion of $\pi = 3,1415926535\ldots$ forms a pseudo-ergodic sequence $\mathbb{N} \to D = \{0, \ldots, 9\}$.

It is easily seen that, if b is pseudo-ergodic with respect to D, then the operator spectrum of $\hat{M}_b$ contains every generalized multiplication operator $\hat{M}_c$ one can think of with a D-valued function c. But also the reverse implication is true.

Proposition 3.68. *Let $\mathbf{X}$ be a Banach space, D be a bounded and closed subset of $L(\mathbf{X})$, and suppose $b \in \ell^\infty(\mathbb{Z}^n, L(\mathbf{X}))$. Then b is pseudo-ergodic with respect to D if and only if*

$$\sigma^{\mathrm{op}}(\hat{M}_b) \supset \{\hat{M}_c : c = (c_\alpha)_{\alpha \in \mathbb{Z}^n} \subset D\}. \qquad (3.33)$$

Proof. In analogy to Proposition 3.27, it is readily seen that, if $A = \hat{M}_b$ with $b = (b_\alpha) \in \ell^\infty(\mathbb{Z}^n, L(\mathbf{X}))$ and $h = (h_m) \subset \mathbb{Z}^n$ tends to infinity, then A_h exists and is equal to $\hat{M}_c$ with $c = (c_\alpha)_{\alpha \in \mathbb{Z}^n} \subset D$ if and only if

$$\sup_{\alpha \in S} \|b_{h_m + \alpha} - c_\alpha\| \to 0 \qquad \text{as} \qquad m \to \infty \qquad (3.34)$$

for all finite sets $S \subset \mathbb{Z}^n$.

Let b be pseudo-ergodic w.r.t. D, and take an arbitrary $c = (c_\alpha)_{\alpha \in \mathbb{Z}^n} \subset D$. For every $m \in \mathbb{N}$, define $h_m \in \mathbb{Z}^n$ as the value of γ in (3.32), where we put

---

[4]Go to http://pi.nersc.gov/ to check if your name is coded in the first 4 billion digits of π.

$\mathbf{Y} = L(\mathbf{X})$, $S = \{-m, \dots, m\}^n$ and $\varepsilon = 1/m$. Since there are infinitely many choices for this vector γ, we can moreover suppose that $|h_m| > m$. Then $h = (h_m)$ converges to infinity, and it is easily seen that (3.34) holds for every bounded $S \subset \mathbb{Z}^n$, showing that $\hat{M}_c \in \sigma^{\mathrm{op}}(\hat{M}_b)$.

Conversely, if $\hat{M}_c \in \sigma^{\mathrm{op}}(\hat{M}_b)$ for every $c = (c_\alpha)_{\alpha \in \mathbb{Z}^n} \subset D$, then (3.34) holds for every finite set $S \subset \mathbb{Z}^n$. But this clearly implies that b is pseudo-ergodic with respect to D. $\qquad\square$

Clearly, for b to be pseudo-ergodic with respect to a set D, it is necessary that D is contained in the closure of the set of all components of b.

Lemma 3.69. *If $b = (b_\alpha) \in \ell^\infty(\mathbb{Z}^n, \mathbf{Y})$ is pseudo-ergodic with respect to $D \subset \mathbf{Y}$, then $D \subset \mathrm{clos}_\mathbf{Y}\{b_\alpha\}_{\alpha \in \mathbb{Z}^n}$.*

Proof. Let b be pseudo-ergodic w.r.t. D, and let $a \in D$ be arbitrary. Put $S = \{0\}$ and $c_0 = a$ to see that, for every $\varepsilon > 0$, there is a $\gamma \in \mathbb{Z}^n$ such that $\|b_\gamma - a\|_\mathbf{Y} < \varepsilon$, by (3.32). $\qquad\square$

We will say that $b = (b_\alpha) \in \ell^\infty(\mathbb{Z}^n, \mathbf{Y})$ is *pseudo-ergodic* if it is pseudo-ergodic with respect to $D = \mathrm{clos}_\mathbf{Y}\{b_\alpha\}_{\alpha \in \mathbb{Z}^n}$, which is the largest possible choice for D.

Corollary 3.70. *Let $\mathbf{X}$ be a Banach space, and suppose $b \in \ell^\infty(\mathbb{Z}^n, L(\mathbf{X}))$. Then b is pseudo-ergodic if and only if*

$$\sigma^{\mathrm{op}}(\hat{M}_b) = \{\hat{M}_c \ : \ c = (c_\alpha)_{\alpha \in \mathbb{Z}^n} \subset D\} \tag{3.35}$$

with $D = \mathrm{clos}_\mathbf{Y}\{b_\alpha\}_{\alpha \in \mathbb{Z}^n}$.

Proof. The only thing that remains to be shown is that the reverse implication holds as well in (3.33). But this follows from Proposition 3.6 d) and the choice of D. $\qquad\square$

Corollary 3.71. *If $b \in \ell^\infty(\mathbb{Z}^n, L(\mathbf{X}))$ is pseudo-ergodic, then $\hat{M}_b$ is among its own limit operators.*

3.4.11 Interplay with Convolution Operators

Remember that the commutator of two operators $A, B \in L(E)$ is the operator $AB - BA$, which is why the operators AB and BA are sometimes referred to as *semi-commutators* of A and B. Moreover, recall from Example 1.45 that the convolution with a L^1-function is a bounded linear operator on every space L^p with $1 \le p \le \infty$.

The Case $1 < p < \infty$

In [81] SHTEINBERG studied the classes $\mathcal{Q}_{\mathrm{SC}}$ and $\mathcal{Q}_{\mathrm{C}}$ of all functions $b \in L^\infty$ for which the semi-commutator or the commutator, respectively, of M_b with an arbitrary L^1-convolution is compact in all spaces L^p with $1 < p < \infty$. This question arises since

- firstly, multiplication and convolution operators are the basic building stones for many interesting band-dominated operators on L^p, as we will see in Chapter 4, for example,

- and secondly, $K(E) \subset K(E, \mathcal{P})$ with $E = L^p$ if $1 < p < \infty$, whence all compact operators on L^p are rich and their operator spectrum is equal to $\{0\}$ if $1 < p < \infty$, by Proposition 3.6 c).

So if $b \in L^\infty$ and C is a L^1-convolution operator on L^p, then

$$\sigma^{\mathrm{op}}(M_b C) \qquad \text{and} \qquad \sigma^{\mathrm{op}}(C M_b) \tag{3.36}$$

are invariant under perturbing b by a function in $\mathcal{Q}_{\mathrm{SC}}$, and

$$\sigma^{\mathrm{op}}(M_b C - C M_b)$$

is invariant under perturbing b by a function in $\mathcal{Q}_{\mathrm{C}}$. The same invariance holds for the answer to the question whether or not the operators $M_b C$, $C M_b$ and $M_b C - C M_b$ are rich, respectively.

It turns out (see [81] or [70, Theorem 3.2.2]) that a function $b \in L^\infty$ is in $\mathcal{Q}_{\mathrm{SC}}$ if and only if, for each compactum $U \subset \mathbb{R}^n$,

$$\left\| b|_{x+U} \right\|_1 \to 0 \qquad \text{as} \qquad x \to \infty, \tag{3.37}$$

where $\|b|_{x+U}\|_1$ denotes the L^1-norm of the restriction of b to the compact set $x + U$. This clearly shows that

$$L_0^\infty \subset \mathcal{Q}_{\mathrm{SC}} \subset \mathcal{Q}_{\mathrm{C}}. \tag{3.38}$$

Moreover (still citing [81] and [70]), $\mathcal{Q}_{\mathrm{C}}$ is a Banach subalgebra of L^∞, $\mathcal{Q}_{\mathrm{SC}}$ is a closed ideal in L^∞, and hence in $\mathcal{Q}_{\mathrm{C}}$, and both algebras are related by the equality

$$\mathcal{Q}_{\mathrm{C}} = \mathcal{Q}_{\mathrm{SC}} + \mathrm{SOC}. \tag{3.39}$$

Actually, we can refine (3.39) a little bit:

Proposition 3.72. $\mathcal{Q}_{\mathrm{C}} = \mathcal{Q}_{\mathrm{SC}} + \mathrm{SO}.$

Proof. From $L_0^\infty \subset \mathcal{Q}_{\mathrm{SC}}$ and both being linear spaces, we get $\mathcal{Q}_{\mathrm{SC}} = \mathcal{Q}_{\mathrm{SC}} + L_0^\infty$. Taking this together with (3.39) and Proposition 3.57, we conclude $\mathcal{Q}_{\mathrm{C}} = \mathcal{Q}_{\mathrm{SC}} + \mathrm{SOC} = \mathcal{Q}_{\mathrm{SC}} + L_0^\infty + \mathrm{SOC} = \mathcal{Q}_{\mathrm{SC}} + \mathrm{SO}$. $\square$

Before we can give some more equivalent characterizations of the algebras Q_{SC} and Q_C, we will have a look at the limit operators of M_b for $b \in Q_{SC}$.

Proposition 3.73. *If* $b \in Q_{SC}$, *then* $\sigma^{op}(M_b)$ *is either* $\{0\}$ *or* $\varnothing$.

Proof. Suppose M_b has a limit operator $(M_b)_h = M_c$. We have to show that $c = 0$. For every compactum $U \subset \mathbb{R}^n$, we have $\|d_m\|_\infty \to 0$ as $m \to \infty$ by (3.18), where $d_m := b|_{h_m+U} - c|_U$. Consequently,

$$
\begin{aligned}
\|c|_U\|_1 &\leq \|b|_{h_m+U}\|_1 + \|d_m\|_1 \\
&\leq \|b|_{h_m+U}\|_1 + |U| \cdot \|d_m\|_\infty \to 0
\end{aligned}
$$

as $m \to \infty$ by (3.37). So $c|_U = 0$ for every compact set $U \subset \mathbb{R}^n$, i.e. $c = 0$. $\square$

Now we are in the position to describe the set of rich functions among Q_{SC} and Q_C, respectively.

Proposition 3.74.   a) $Q_{SC} \cap L_\$^\infty = L_0^\infty$.

 b) $Q_C \cap L_\$^\infty = SO$.

Proof. If $b \in Q_{SC}$ and $b \in L_\$^\infty$, then $\sigma^{op}(M_b) = \{0\}$ by Proposition 3.73, and Proposition 3.54 tells that $b \in L_0^\infty$. The reverse inclusion is trivial.

If $f \in Q_C \cap L_\$^\infty$, then, by Proposition 3.72, we get $f = g+h$ with $g \in Q_{SC}$ and $h \in SO \subset L_\$^\infty$, by Proposition 3.49. Consequently, $f \in L_\$^\infty$ implies $g = f-h \in L_\$^\infty$, and from **a)**, we get $g \in L_0^\infty \subset SO$. So $g, h \in SO$, whence $f = g + h \in SO$. For the reverse inclusion, recall Propositions 3.72 and 3.49. $\square$

Having this result, we will try to find alternative descriptions of Q_{SC} and Q_C. We already noticed that Q_{SC} contains L_0^∞. But it is strictly larger since it also contains functions b with property (3.37) and

$$
\big\| b|_{x+U} \big\|_\infty \not\to 0 \qquad \text{as } x \to \infty. \tag{3.40}
$$

We will refer to functions which are subject to (3.37) and (3.40) as *noise*. A typical example of a noise function is the characteristic function of the set

$$
\bigcup_{m=1}^\infty \left[m, \, m + \frac{1}{m} \right] \subset \mathbb{R}
$$

for $n = 1$. The set of all noise functions shall be denoted by $\mathcal{N}$.

If $A \cup B$ denotes the union of two disjoint sets A, B, then the following holds.

Lemma 3.75. $Q_{SC} = L_0^\infty \cup \mathcal{N}$.

Proof. This is trivial: The set of functions subject to (3.37) decomposes into those with (3.37) and (3.40), which is $\mathcal{N}$, and those with (3.37) and not (3.40), which is L_0^∞. $\square$

Corollary 3.76. *Noise is never rich. Even shorter,* $\mathcal{N} \cap L_\$^\infty = \varnothing$.

Proof 1. Use Lemma 3.75 and Proposition 3.74 a).   □

Proof 2. Conversely, suppose $b \in \mathcal{N}$ is rich. Take an integer sequence $h = (h_m)$ tending to infinity with $\|b|_{h_m+Q}\|_\infty$ being bounded away from zero. But from Proposition 3.73 we know that $(M_b)_h = 0$, which clearly contradicts (3.18).   □

For a concise description of $\mathcal{Q}_{SC}$ and $\mathcal{Q}_C$, put $\mathcal{N}_0 := \mathcal{N} \cup \{0\}$.

Proposition 3.77. *Both $\mathcal{Q}_{SC}$ and $\mathcal{Q}_C$ decompose into a rich and a noisy part by $\cup$ and by $+$:*

> a) $\mathcal{Q}_{SC} = L_0^\infty \cup \mathcal{N}$ *and*   b) $\mathcal{Q}_C = SO \cup (SO + \mathcal{N})$.
> c) $\mathcal{Q}_{SC} = L_0^\infty + \mathcal{N}_0$ *and*   d) $\mathcal{Q}_C = SO + \mathcal{N}_0$.

Proof. a) is a repetition of Lemma 3.75.

Considering b), we recall Proposition 3.72, part **a)** and $L_0^\infty \subset SO$ to get

$$\mathcal{Q}_C = \mathcal{Q}_{SC} + SO = (L_0^\infty \cup \mathcal{N}) + SO = (L_0^\infty + SO) \cup (\mathcal{N} + SO) = SO \cup (\mathcal{N} + SO).$$

Clearly, SO and $SO + \mathcal{N}$ are disjoint since the earlier functions are always rich by Proposition 3.49, and the latter are always ordinary by Proposition 3.49 and Corollary 3.76.

Now c) follows from a) and $\mathcal{N} = \mathcal{N} + L_0^\infty$ by

$$\mathcal{Q}_{SC} = L_0^\infty \cup \mathcal{N} = L_0^\infty \cup (L_0^\infty + \mathcal{N}) = L_0^\infty + (\mathcal{N} \cup \{0\}),$$

and d) follows from b) by

$$\mathcal{Q}_C = (SO + \mathcal{N}) \cup SO = SO + (\mathcal{N} \cup \{0\}).$$ □

We conclude this journey by a simple corollary.

Corollary 3.78.   a) *If $b \in \mathcal{Q}_{SC}$ and $(M_b)_h = M_c$ exists, then $c = 0$.*

   b) *If $b \in \mathcal{Q}_C$ and $(M_b)_h = M_c$ exists, then $c = const$.*

Proof. a) is a repetition of Proposition 3.73, and b) follows from Proposition 3.77 b) and Corollary 3.48.   □

The Cases $p = 1$ and $p = \infty$

We will see that, if $p = 1$ or $p = \infty$, one of the two operator spectra (3.36) as well as the richness of the corresponding operator is still invariant under perturbing b by a function in $\mathcal{Q}_{SC}$.

Recall that, for $b \in L^\infty$, the multiplication operator M_b is rich if every sequence in $\mathbb{Z}^n$ tending to infinity has an infinite subsequence $h = (h_m)$ such that there exists a function $c \in L^\infty$ with

$$\| b|_{h_m + U} - c|_U \|_\infty \to 0 \quad \text{as} \quad m \to \infty \tag{3.41}$$

for every compactum $U \subset \mathbb{R}^n$, using the lazy notation as in Remark 3.26. We will show that the semi-commutator of M_b with a convolution operator is already rich if the above holds with the much weaker condition

$$\left\| b|_{h_m + U} - c|_U \right\|_1 \to 0 \quad \text{as} \quad m \to \infty \tag{3.42}$$

instead of (3.41). We denote the set of all $b \in L^\infty$ with this property by $L^\infty_{\text{SC\$}}$, and write $\tilde{b}^{(h)}$ for the function c with property (3.42) for all compact sets U. Recall that $L^\infty_{\$}$ is the set of all $b \in L^\infty$ with the first property, and that we write $b^{(h)}$ for the function c in (3.41).

Proposition 3.79. *Let C be the operator of convolution on L^p by a function $\kappa \in L^1$, let $b \in L^\infty_{\text{SC\$}}$, and $h \subset \mathbb{Z}^n$ be a sequence tending to infinity.*

  a) *If $p = 1$, then $M_b C$ is rich and $(M_b C)_h = M_{\tilde{b}^{(h)}} C$ if $\tilde{b}^{(h)}$ exists.*

  b) *If $p = \infty$, then $C M_b$ is rich and $(C M_b)_h = C M_{\tilde{b}^{(h)}}$ if $\tilde{b}^{(h)}$ exists.*

Proof. Since κ can be approximated in the norm of L^1 as close as desired by a continuous function with compact support, it is sufficient to prove the proposition for this case. Choose $\ell > 0$ large enough for $\operatorname{supp} \kappa \subset [-\ell, \ell]^n$.

   We prove the statement b). Then statement a) follows from duality. To see that $C M_b$ is rich, take an arbitrary sequence $g \subset \mathbb{Z}^n$ tending to infinity. Since $b \in L^\infty_{\text{SC\$}}$, there is an infinite subsequence $h = (h_m)$ of g and a function $\tilde{c} = b^{(h)} \in L^\infty$ such that (3.42) holds for all compact sets $U \subset \mathbb{R}^n$. Then, for every $\tau > 0$, putting $U := [-\tau - \ell, \tau + \ell]^n$,

$$\| P_\tau (V_{-h_m} C M_b V_{h_m} - C M_{\tilde{b}^{(h)}}) \| \leq \operatorname*{ess\,sup}_{|x| < \tau} \int_{\mathbb{R}^n} |\kappa(x-y)| \, |b(y + h_m) - \tilde{b}^{(h)}(y)| \, dy$$

$$= \operatorname*{ess\,sup}_{|x| < \tau} \int_U |\kappa(x-y)| \, |b(y + h_m) - \tilde{b}^{(h)}(y)| \, dy$$

$$\leq \|\kappa\|_\infty \left\| b|_{h_m + U} - \tilde{b}^{(h)}|_U \right\|_1 \to 0$$

as $m \to \infty$. Analogously, for every $\tau > 0$, putting $V = [-\tau, \tau]^n$,

$$\| (V_{-h_m} C M_b V_{h_m} - C M_{\tilde{b}^{(h)}}) P_\tau \| \leq \operatorname*{ess\,sup}_{x \in \mathbb{R}^n} \int_V |\kappa(x-y)| \, |b(y + h_m) - \tilde{b}^{(h)}(y)| \, dy$$

$$\leq \|\kappa\|_\infty \left\| b|_{h_m + V} - \tilde{b}^{(h)}|_V \right\|_1 \to 0$$

as $m \to \infty$. This shows that $C M_{\tilde{b}^{(h)}}$ is the limit operator of $C M_b$ with respect to the subsequence h of g. $\qquad\square$

Proposition 3.80. *It holds that $L^\infty_{\$} + \mathcal{Q}_{\text{SC}} = L^\infty_{\$} + \mathcal{N}_0 \subset L^\infty_{\text{SC\$}}$.*

Proof. From Proposition 3.77 c) we get that $L^\infty_{\$} + \mathcal{Q}_{\text{SC}} = L^\infty_{\$} + L^\infty_0 + \mathcal{N}_0 = L^\infty_{\$} + \mathcal{N}_0$. Now let $b = c + d$ with $c \in L^\infty_{\$}$ and $d \in \mathcal{Q}_{\text{SC}}$, and let $g \subset \mathbb{Z}^n$ be an arbitrary sequence tending to infinity. Since $c \in L^\infty_\$$, there is a subsequence $h = (h_m) \subset g$ and a function $\tilde{c}^{(h)}$ such that $\| c|_{h_m+U} - \tilde{c}^{(h)}|_U \|_\infty \to 0$ as $m \to \infty$ for every compact set U in $\mathbb{R}^n$. But then

$$
\begin{aligned}
\left\| b|_{h_m+U} - \tilde{c}^{(h)}|_U \right\|_1 &\leq \left\| c|_{h_m+U} - \tilde{c}^{(h)}|_U \right\|_1 + \left\| d|_{h_m+U} \right\|_1 \\
&\leq |U| \cdot \left\| c|_{h_m+U} - \tilde{c}^{(h)}|_U \right\|_\infty + \left\| d|_{h_m+U} \right\|_1 \to 0
\end{aligned}
$$

as $m \to \infty$ for every compact set U in $\mathbb{R}^n$. $\square$

Corollary 3.81. *Let C be the operator of convolution on L^p by a function in L^1, and let $b \in L^\infty$. If $p = 1$, then the richness and the operator spectrum of $M_b C$ are invariant under perturbations of b in $\mathcal{Q}_{SC}$. If $p = \infty$, then the richness and the operator spectrum of $C M_b$ are invariant under perturbations of b in $\mathcal{Q}_{SC}$.*

3.4.12   The Big Picture

We haven't had a picture for quite a long time. So let us look at another Venn diagram that shows the relation of most of the classes of functions studied in the previous subsections. Here is a short reminder and a legend to the picture.

$L^\infty_\$$	...	the rich functions;
BC	...	the bounded and continuous functions;
BUC	=	$BC \cap L^\infty_\$$, the bounded and uniformly continuous functions;
CL	...	the functions $b \in L^\infty$ for which $\sigma^{op}(M_b) \subset \{cI : c \in \mathbb{C}\}$;
SO	=	$CL \cap L^\infty_\$$, the slowly oscillating functions;
SOC	=	$SO \cap BC$, the slowly oscillating continuous functions;
$\mathcal{Q}_C$	...	the functions b for which $[M_b, C] \in K(L^p)$ for all $p \in (1, \infty)$ and all convolutions C by a L^1-function;
$\mathcal{Q}_{SC}$	...	the functions b for which $M_b C, C M_b \in K(L^p)$ for all $p \in (1, \infty)$ and all convolutions C by a L^1-function;
L^∞_0	=	$\mathcal{Q}_{SC} \cap L^\infty_\$$, the functions vanishing at infinity;
$\mathcal{N}$	=	$\mathcal{Q}_{SC} \cap (L^\infty \setminus L^\infty_\$)$, the noise functions;
CL_1	...	the functions $b \in CL$ for which $\sigma^{op}_s(M_b)$ has at most 1 element per direction $s \in S^{n-1}$;
AP	...	the almost periodic functions;
SAP	...	the semi-almost periodic functions.

Moreover, we have the equalities

$$
\begin{aligned}
C(\overline{\mathbb{R}^n}) + L^\infty_0 &= CL_1 \cap L^\infty_\$, \\
C(\overline{\mathbb{R}^n}) &= CL_1 \cap L^\infty_\$ \cap BC, \\
SO &= \mathcal{Q}_C \cap L^\infty_\$.
\end{aligned}
$$

Venn Diagram for Some Subclasses of L^∞

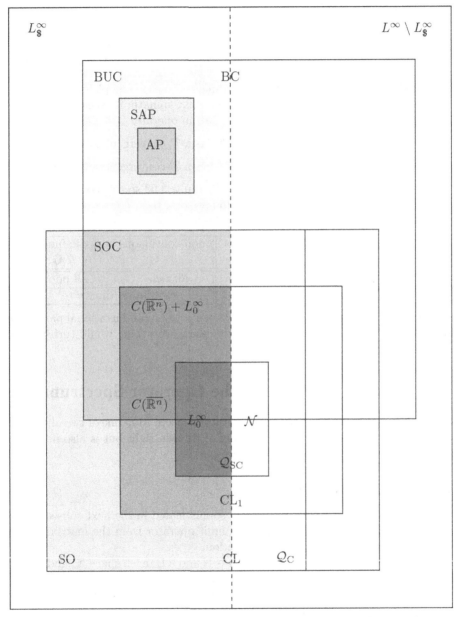

Figure 8: The relationship between the subclasses of L^∞ studied in Section 3.4.

Note that, for the sake of simplicity of the picture, we have excluded the constant functions from AP and SAP for a moment.

3.4.13   Why Choose Integer Sequences?

In Subsections 3.4.2, 3.4.8 and 3.4.9 we have experienced the consequences of the restriction of $h = (h_m)$ to integer sequences, as already discussed in Remark 3.28 b). We have seen that step functions with rational step length and periodic functions with rational period are always rich while their irrational counterparts are ordinary or even poor in general. This seems a bit unnatural since re-scaling axes a little bit will change the rich-or-ordinary-situation completely.

The reason why we stick to this somewhat unnatural constraint becomes apparent when we look at what it means that an operator $A \in L(E)$ is rich:

A: Every sequence (3.5) with $(h_m) \subset \mathbb{Z}^n$ has a $\mathcal{P}$-convergent subsequence.

B: Every sequence (3.5) with $(h_m) \subset \mathbb{R}^n$ has a $\mathcal{P}$-convergent subsequence.

Clearly, **B** implies **A**, but, as already Example 3.63 shows, not the other way round. By passing from $\mathbb{Z}^n$ to $\mathbb{R}^n$, we would lose some basic classes of rich functions frequently used in applications!

	non-convergent step functions		non-continuous periodic functions	
	$s \in \mathbb{Q}$	$s \notin \mathbb{Q}$	$\omega \in \mathbb{Q}$	$\omega \notin \mathbb{Q}$
$h \subset \mathbb{Z}$	all rich	ordinary/poor	all rich	all poor
$h \subset \mathbb{R}$	all ordinary		all ordinary	

Indeed, the unnatural distinction between rational and irrational parameters would vanish. But instead of these distinct cases, everything would turn into one case of ordinary functions.

3.5   Alternative Views on the Operator Spectrum

In this section we will look at the limit operator concept from one or two alternative perspectives, which is certainly interesting in its own right but is also helpful for a better understanding of the whole concept.

3.5.1   The Matrix Point of View

Before we come to something much more sophisticated in the next subsection, we first have to think about the notion of a limit operator from the matrix point of view, as it was motivated in the introduction.

Therefore take some $p \in [1, \infty]$, an $n \in \mathbb{N}$ and a Banach space $\mathbf{X}$, and remember that with every bounded linear operator A on $E = \ell^p(\mathbb{Z}^n, \mathbf{X})$ we associated a matrix $[A]$ with entries $a_{\alpha\beta} \in L(\mathbf{X})$, indexed over $\mathbb{Z}^n \times \mathbb{Z}^n$. If A is band-dominated, then also the matrix $[A]$ uniquely determines the operator A.

An elementary computation shows that, for every $c \in \mathbb{Z}^n$, the matrix $[b_{\alpha\beta}]$ of the shifted operator $V_{-c} A V_c$ is connected with $[A] = [a_{\alpha\beta}]$ by

$$b_{\alpha,\beta} = a_{\alpha+c,\beta+c}, \qquad \alpha, \beta \in \mathbb{Z}^n;$$

that is, roughly speaking, $[V_{-c}AV_c]$ is what you see when you look at $[A]$ from the point that is c positions further down the main diagonal.

Now we know what the matrices $[V_{-h_m}AV_{h_m}]$ look like when $[A]$ is given and $h = (h_m) \subset \mathbb{Z}^n$ is a sequence that tends to infinity. To go one step further and get an idea what $[A_h]$ looks like, provided the limit operator A_h exists, we have to understand $\mathcal{P}$-convergence from the matrix point of view.

Therefore take a bounded sequence $(A_m) \subset L(E)$ and an $A \in L(E)$. From Proposition 1.65 we easily derive that $A_m \xrightarrow{\mathcal{P}} A$ if and only if, for all $\gamma \in \mathbb{Z}^n$,

$$P_\gamma A_m \rightrightarrows P_\gamma A \quad \text{and} \quad A_m P_\gamma \rightrightarrows A P_\gamma \quad \text{as} \quad m \to \infty;$$

that is, every row and every column of A_m uniformly converges to the respective row and column of A. In this connection also recall that every row and every column of a matrix decays towards infinity if the underlying operator is in $L(E, \mathcal{P})$.

Summarizing, this is how the process of passing to the limit operator for a given sequence $h = (h_1, h_2, \ldots)$ looks like from the matrix point of view:

- Travel down the main diagonal of $[A]$ along the sites $h_1, h_2, \ldots$, where, at each site, you look at the matrix through an infinite cross-shaped window that only allows to view at a finite number k of rows and columns.

- If, during this journey down the diagonal, uniform convergence happens in your window, and if this is the case for all widths k of the cross bars, then $h = (h_m)$ leads to a limit operator of A, and the rows and columns of $[A_h]$ are the uniform limits of our respective outlooks.

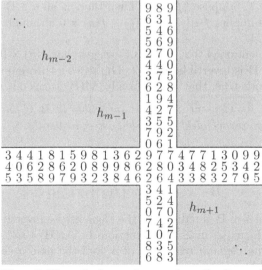

Figure 9: The outlook on a journey down the diagonal while looking through a cross-shaped window with $k = 3$.

3.5.2   Another Parametrization of the Operator Spectrum

Instead of the long journey down the diagonal, waiting for convergence, one can give a much more straightforward definition for every matrix entry of a limit operator. Before we can give this definition, we need some more knowledge on commutative Banach algebras.

Maximal Ideals and Characters

Let $\mathbf{B}$ be a commutative Banach algebra with unit e. A function $f : \mathbf{B} \to \mathbb{C}$ which is subject to $f(a \dotplus b) = f(a) \dotplus f(b)$ for all $a, b \in \mathbf{B}$ and which is not the zero function on $\mathbf{B}$ is called a *character* of $\mathbf{B}$. Characters are sometimes also called states, non-zero scalar-valued homomorphisms or non-zero multiplicative functionals. Moreover, an ideal $M \subset \mathbf{B}$ is a *proper ideal* of $\mathbf{B}$ if $M \neq \mathbf{B}$, and it is called *maximal ideal* of $\mathbf{B}$ if it is a proper ideal of $\mathbf{B}$ that is not properly contained in another proper ideal of $\mathbf{B}$. The family of all maximal ideals M of a commutative Banach algebra $\mathbf{B}$ is denoted by $\mathcal{M}(\mathbf{B})$.

There is a one-to-one correspondence between characters and maximal ideals of a commutative Banach algebra $\mathbf{B}$. The kernel of every character is a maximal ideal, and conversely, every maximal ideal is the kernel of a uniquely determined character. Therefore, one often thinks of the elements of $\mathcal{M}(\mathbf{B})$ as characters and writes $f \in \mathcal{M}(\mathbf{B})$ if f is a character of $\mathbf{B}$.

It is clear that every character of $\mathbf{B}$ maps the unit e to 1, and, by (3.1), every invertible element in $\mathbf{B}$ is mapped to a non-zero complex number. As a consequence of this, we get that every character is automatically bounded as a linear functional, and its norm is equal to 1. Indeed, let $f \in \mathcal{M}(\mathbf{B})$. From $f(e) = 1$ we get that $\|f\| \geq 1$. Suppose $\|f\| > 1$. Then there is an $a \in \mathbf{B}$ with $\|a\| < 1$ and $f(a) = 1$. But this implies $f(e - a) = f(e) - f(a) = 0$ although $e - a$ is invertible in $\mathbf{B}$ by Lemma 1.2.

We already mentioned that from (3.1) we get that $f(a) \neq 0$ for all characters $f \in \mathcal{M}(\mathbf{B})$ if $a \in \mathbf{B}$ is invertible. In fact, GELFAND's theorem says that also the reverse implication is true; that is, the family $\mathcal{M}(\mathbf{B})$ turns out to be sufficient.

Proposition 3.82. *An element $a \in \mathbf{B}$ is invertible in $\mathbf{B}$ if and only if it is not contained in any maximal ideal of $\mathbf{B}$; that is, equivalently, if no character of $\mathbf{B}$ vanishes on a.*

This observation yields the finest[5] possible localization of an element in $\mathbf{B}$:

$$a \in \mathbf{B} \qquad \longleftrightarrow \qquad \{f(a)\}_{f \in \mathcal{M}(\mathbf{B})} \subset \mathbb{C}$$

The invertibility of a in $\mathbf{B}$ is reduced to the elementwise invertibility of the family of numbers $\{f(a)\}_{f \in \mathcal{M}(\mathbf{B})}$ in $\mathbb{C}$. The function $\hat{a} : \mathcal{M}(\mathbf{B}) \to \mathbb{C}$ with $\hat{a}(f) = f(a)$ is referred to as the *Gelfand transform* of a, and the map $a \mapsto \hat{a}$ is the *Gelfand transformation* on $\mathbf{B}$.

---

[5]One clearly cannot 'decompose' an $a \in \mathbf{B}$ into something 'finer' than complex numbers.

The set $\mathcal{M}(\mathbf{B})$ is a compact Hausdorff space under the Gelfand topology – the coarsest topology on $\mathcal{M}(\mathbf{B})$ that makes all Gelfand transforms $\hat{a} : \mathcal{M}(\mathbf{B}) \to \mathbb{C}$ with $a \in \mathbf{B}$ continuous.

Example 3.83. In Example 3.1, where $\mathbf{B} = C[0,1]$, the evaluation functionals

$$\varphi_x : a \mapsto a(x), \qquad a \in \mathbf{B}$$

with $x \in [0,1]$ are exactly the characters of $\mathbf{B}$, so that $\mathcal{M}(\mathbf{B})$ can be identified with $[0,1]$ by identifying φ_x with x. After this identification, the Gelfand transform $\hat{a} : \mathcal{M}(\mathbf{B}) \to \mathbb{C}$ coincides with the function $a : [0,1] \to \mathbb{C}$ itself, and the Gelfand topology on $\mathcal{M}(\mathbf{B})$ coincides with the usual topology on $[0,1]$. $\qquad\square$

Characters of ℓ^∞

For our purposes, we are especially interested in the characters of the commutative Banach algebra $\mathbf{B} = \ell^\infty = \ell^\infty(\mathbb{Z}^n, \mathbb{C})$ with unit element $e = (1)_{\alpha \in \mathbb{Z}^n}$. The set $\mathcal{M}(\ell^\infty)$ naturally subdivides in two classes:

- Characters that evaluate $a \in \ell^\infty$ at finite points.

- Characters that evaluate $a \in \ell^\infty$ at infinity.

To see this, introduce the subspace ℓ_0^∞ of all elements $a \in \ell^\infty$ that decay at infinity. A character $f \in \mathcal{M}(\ell^\infty)$ is in the class $\mathcal{M}_\infty(\ell^\infty)$ if f vanishes on ℓ_0^∞, otherwise it is in the class $\mathcal{M}_{\mathbb{Z}^n}(\ell^\infty)$. Equivalently, a maximal ideal M of ℓ^∞ is in $\mathcal{M}_\infty(\ell^\infty)$ if and only if it is contained in ℓ_0^∞. As a closed subset of $\mathcal{M}(\ell^\infty)$, the set $\mathcal{M}_\infty(\ell^\infty)$ is compact in the Gelfand topology as well.

It is an elementary exercise to show that $f \in \mathcal{M}_{\mathbb{Z}^n}(\ell^\infty)$ if and only if f is an evaluation functional; that is

$$f(a) = a_\beta, \qquad a = (a_\alpha)_{\alpha \in \mathbb{Z}^n} \in \ell^\infty$$

for some fixed $\beta \in \mathbb{Z}^n$. In this sense, the class $\mathcal{M}_{\mathbb{Z}^n}(\ell^\infty)$ can be identified with the set $\mathbb{Z}^n$. The much more interesting class is $\mathcal{M}_\infty(\ell^\infty)$. In contrast to characters in $\mathcal{M}_{\mathbb{Z}^n}(\ell^\infty)$, it is convenient to think of elements of $\mathcal{M}_\infty(\ell^\infty)$ as evaluation functionals at infinity.

General elements of ℓ^∞ may behave pretty arbitrarily at infinity which makes it a hard task to actually think of a character $f \in \mathcal{M}_\infty(\ell^\infty)$ that evaluates a particular aspect of this behaviour with a single number. If we pass to the subalgebra ℓ_c^∞ of ℓ^∞ that consists of all convergent sequences $a = (a_\alpha) \in \ell^\infty$ with $a_\alpha \to a_\infty$ as $\alpha \to \infty$, a character of ℓ_c^∞ acting at infinity is certainly the functional $f_{\lim}$ that assigns to every element $a \in \ell_c^\infty$ its limit a_∞ at infinity.

Although this limit functional $f_{\lim}$ cannot be defined on all of ℓ^∞, it is nice to see that every character f of ℓ^∞ acting at infinity is just a continuation[6] of

---

[6]At this point you might also want to recall the functional in Example 1.26 c). But remember that the characters $f \in \mathcal{M}_\infty(\ell^\infty)$, on top of that, also have to be multiplicative.

$f_{\lim}$ from ℓ_c^∞ to all of ℓ^∞. Or, the other way round, any character of ℓ^∞ acting at infinity, applied to a convergent sequence $a \in \ell_c^\infty$, returns nothing but the obvious value $f_{\lim}(a) = a_\infty$. Indeed, if $a \in \ell_c^\infty$, then a can be written as $a = a_\infty e + c$ with $c \in \ell_0^\infty$, whence

$$f(a) = a_\infty f(e) + f(c) = a_\infty \cdot 1 + 0 = f_{\lim}(a)$$

for all $f \in \mathcal{M}_\infty(\ell^\infty)$.

The Stone-Čech Boundary

A slight generalization of the previous notions is the following. If G is an infinite discrete group, then the set

$$\beta G := \mathcal{M}(\ell^\infty(G))$$

is called the *Stone-Čech compactification of* G. As in the previous case, $G = \mathbb{Z}^n$, the set βG subdivides into a set of characters acting at infinity and another set of characters acting at the points of G. The latter set can again be identified with G itself, and the first set is called the *Stone-Čech boundary* of G, denoted by

$$\partial G := \beta G \setminus G.$$

So in our case, $G = \mathbb{Z}^n$, we have $\beta \mathbb{Z}^n = \mathcal{M}(\ell^\infty)$ and $\partial \mathbb{Z}^n = \mathcal{M}_\infty(\ell^\infty)$.

The Stone-Čech compactification βG has the so-called universal property: Every map from G to a compact Hausdorff space K can be uniquely extended to a continuous map $\beta G \to K$ and hence to the boundary ∂G.

Rearranging the Operator Spectrum

Remember that we motivated the introduction of limit operators by the idea of studying an invertibility problem via a family of unital homomorphisms. But although we finally proved Theorem 1, our operator spectrum has a little defect:

For a fixed sequence $h \subset \mathbb{Z}^n$ tending to infinity, the mapping $A \mapsto A_h$ cannot be defined on all of $L(E)$ – not even on $L(E, \mathcal{P})$ or BDO^p. The best thing one can do is to study the restriction of $A_h \mapsto A$ to the set $\{A \in \text{BDO}^p : A_h \text{ exists}\}$, which is a Banach algebra by Proposition 3.4, where it is indeed a homomorphism, by the same Proposition 3.4. But this is not very satisfactory.

Alternatively, instead of fixing h, we could first fix A, denote the set of all sequences $h = (h_m) \subset \mathbb{Z}^n$ with $h_m \to \infty$ for which A_h exists by $\mathcal{H}_A$ and then study the map

$$A \mapsto \sigma^{\text{op}}(A) = \{A_h\}_{h \in \mathcal{H}_A},$$

hoping that this is a homomorphism from $L(E, \mathcal{P})$ or BDO^p to an algebra of operator-valued functions on $\mathcal{H}_A$. But of course there is the same problem as above since $\mathcal{H}_A$ heavily depends on A. This defect prevents us from writing things like

$$\sigma^{\mathrm{op}}(A+B) = \sigma^{\mathrm{op}}(A) + \sigma^{\mathrm{op}}(B) \qquad \text{and} \qquad \sigma^{\mathrm{op}}(AB) = \sigma^{\mathrm{op}}(A)\sigma^{\mathrm{op}}(B) \quad (3.43)$$

in the pointwise sense, since $\sigma^{\mathrm{op}}(A)$ and $\sigma^{\mathrm{op}}(B)$ are defined on the, in general different, sets $\mathcal{H}_A$ and $\mathcal{H}_B$. And even if $\mathcal{H}_A = \mathcal{H}_B$, the sets $\mathcal{H}_{A+B}$ and $\mathcal{H}_{AB}$ can be very different from those, as simple examples like $B = -A$ show.

Summarizing this little dilemma, Theorem 1 shows that our operator spectrum $\sigma^{\mathrm{op}}(A)$ contains the **right elements**, but, as the above problems unveil, with a **slightly weird enumeration**. The solution to these problems is a rearrangement of the operator spectrum as

$$\sigma^{\mathrm{op}}(A) = \{A_f\}_{f \in F}$$

for every $A \in \mathrm{BDO}^p$ in a way that the index set F is no longer dependent on A. The answer, presented in [67], is to choose $F = M_\infty(\ell^\infty)$ and to define A_f in an appropriate way for every $f \in M_\infty(\ell^\infty)$.

Therefore we restrict ourselves to $E = \ell^p = \ell^p(\mathbb{Z}^n)$ with $p \in [1, \infty]$; that is, we put $\mathbf{X} = \mathbb{C}$. First suppose $A \in L(E)$ is a band operator, so we can write A in the form (1.18),

$$A = \sum_{|\gamma| \le w} \hat{M}_{b^{(\gamma)}} V_\gamma$$

with $w \in \mathbb{N}_0$ and $b^{(\gamma)} \in \ell^\infty$ for every γ under consideration. Instead of taking the long journey down the diagonal of A and wait for convergence as described in Subsection 3.5.1, we will immediately evaluate the diagonals $b^{(\gamma)} \in \ell^\infty$ at infinity, using a character $f \in M_\infty(\ell^\infty)$. This gives us one number for every diagonal $b^{(\gamma)}$ of A, which will be an entry of the respective diagonal $c^{(\gamma)}$ of the limit operator A_f. The other entries of $c^{(\gamma)}$ are derived in the same way by considering shifts of A and hence shifts of the diagonal $b^{(\gamma)}$. So put

$$A_f := \sum_{|\gamma| \le w} \hat{M}_{c^{(\gamma)}} V_\gamma$$

with $c^{(\gamma)} = \left(f(V_{-\alpha} b^{(\gamma)}) \right)_{\alpha \in \mathbb{Z}^n} \in \ell^\infty$ for every $\gamma \in \mathbb{Z}^n$ with $|\gamma| \le w$; that is

$$[A_f] = \begin{pmatrix} \ddots & & \ddots & & \ddots & & \ddots & & 0 \\ & \ddots & f(V_1 b^{(0)}) & f(V_1 b^{(-1)}) & f(V_1 b^{(-2)}) & \ddots \\ & \ddots & f(b^{(1)}) & f(b^{(0)}) & f(b^{(-1)}) & \ddots \\ & \ddots & f(V_{-1} b^{(2)}) & f(V_{-1} b^{(1)}) & f(V_{-1} b^{(0)}) & \ddots \\ 0 & & \ddots & & \ddots & & \ddots & & \ddots \end{pmatrix}$$

as an illustration for the case $n = 1$.

Remark 3.84. As a trivial consequence of the formula for $c^{(\gamma)}$, we see that, if a diagonal $b^{(\gamma)}$ of A converges to a value b_∞ at infinity, then the respective diagonal $c^{(\gamma)}$ of every limit operator A_f is constantly equal to b_∞. $\qquad\qquad\square$

In Proposition 20 of [67] it is shown that, for every $f \in \mathcal{M}_\infty(\ell^\infty)$, the map $A \mapsto A_f$ is a bounded homomorphism on all of BO, and hence it can be extended to a bounded homomorphism on BDO^p. Moreover, in Theorem 5 of [67] it is shown that

$$\{A_f\}_{f \in \mathcal{M}_\infty(\ell^\infty)} = \sigma^{\mathrm{op}}(A) = \{A_h\}_{h \in \mathcal{H}_A}$$

holds for every $A \in \mathrm{BDO}^p$. With this new enumeration, we can identify $\sigma^{\mathrm{op}}(A)$ with a function $\mathcal{M}_\infty(\ell^\infty) \to \mathrm{BDO}^p$ mapping $f \mapsto A_f$, for which (3.43) holds by the results stated above.

In [73] ROE reformulates this idea in terms of the Stone-Čech compactification $\partial\mathbb{Z}^n$ for the C^*-algebra case $p = 2$. Recall from Figures 1, 3 and 4 on pages 15, 46 and 57, respectively, that, in this case, $K(E,\mathcal{P}) = K(E)$, $\mathcal{P}$-convergence equals $*$-strong convergence, and invertibility at infinity equals Fredholmness. The elegant approach of [73] even allows the substitution of $\mathbb{Z}^n$ by any finitely generated discrete group. The plot is basically as follows.

Fix a band-dominated operator A on $E = \ell^2$. By the universal property, there is a uniquely determined $*$-strongly continuous extension of the function

$$\mathbb{Z}^n \to \mathrm{BDO}^2 \qquad \text{mapping} \qquad \gamma \mapsto V_{-\gamma} A V_\gamma$$

to the Stone-Čech compactification $\beta\mathbb{Z}^n$. The operator spectrum $\sigma^{\mathrm{op}}(A)$ is then defined as the restriction of this function to the Stone-Čech boundary $\partial\mathbb{Z}^n$. Consequently, $\sigma^{\mathrm{op}}(A)$ is an element of the C^*-algebra C_s of all $*$-strongly continuous functions $\partial\mathbb{Z}^n \to \mathrm{BDO}^2$, equipped with the supremum norm. Finally, the map

$$\sigma^{\mathrm{op}} : A \mapsto \sigma^{\mathrm{op}}(A)$$

is a $*$-homomorphism from BDO^2 to this algebra C_s, the kernel of which coincides with the set of compact operators $K(E) = K(E,\mathcal{P})$. These facts immediately show that $\mathrm{BDO}^2/K(E,\mathcal{P})$ is $*$-isomorphic to the image of the $*$-homomorphism σ^{op} in C_s, proving Theorem 1 for the case $p = 2$, $\mathbf{X} = \mathbb{C}$.

The Local Principle by ALLAN and DOUGLAS

For commutative Banach algebras $\mathbf{B}$, the Gelfand transform $\hat{a}$ yields a very fine localization of an element $a \in \mathbf{B}$. A useful localization technique for non-commutative Banach algebras is the local principle by ALLAN and DOUGLAS, also known as *central localization*.

The set of all $a \in \mathbf{B}$ that commute with any element of $\mathbf{B}$ is referred to as the *center*, denoted by $\mathrm{cen}\,\mathbf{B}$, of $\mathbf{B}$. Clearly, $\mathrm{cen}\,\mathbf{B}$ is a commutative Banach subalgebra of $\mathbf{B}$ which coincides with $\mathbf{B}$ if and only if $\mathbf{B}$ is commutative. In either case, $\mathrm{cen}\,\mathbf{B}$ contains at least the multiples of the unit; that is $\mathbb{C}e = \{ce : c \in \mathbb{C}\}$. In the case $\operatorname{cen} \mathbf{B} = \mathbb{C}e$, we say that $\mathbf{B}$ only has the trivial center. So, in a sense, the center of $\mathbf{B}$ measures the amount of commutativity in $\mathbf{B}$. The central localization works as follows.

Fix a Banach subalgebra $\mathbf{A}$ of $\operatorname{cen} \mathbf{B}$ containing the unit e, which makes $\mathbf{A}$ a commutative Banach algebra. For every maximal ideal $M \in \mathcal{M}(\mathbf{A})$, denote by $J_M := \operatorname{closid}_{\mathbf{B}} M$ the smallest ideal in $\mathbf{B}$ that contains M. Moreover, abbreviate the so-called *local algebra* $\mathbf{B}/J_M$ by $\mathbf{B}_M$, and write b_M for the so-called *local representatives* $b + J_M \in \mathbf{B}_M$ of $b \in \mathbf{B}$. The local principle of ALLAN [1] and DOUGLAS [27] then says the following.

Proposition 3.85. *An element $b \in \mathbf{B}$ is invertible in $\mathbf{B}$ if and only if, for all $M \in \mathcal{M}(\mathbf{A})$, the local representative b_M is invertible in its local algebra $\mathbf{B}_M$.*

For a commutative Banach algebra $\mathbf{B}$, one can choose $\mathbf{A} = \operatorname{cen} \mathbf{B} = \mathbf{B}$. But then, for every $M \in \mathcal{M}(\mathbf{A}) = \mathcal{M}(\mathbf{B})$, we get that $J_M = M$ is the kernel of a character f_M of $\mathbf{B}$. So the local algebra $\mathbf{B}_M = \mathbf{B}/J_M = \mathbf{B}/\ker f_M$ is isomorphic to $\operatorname{im} f_M = \mathbb{C}$. In this case, Proposition 3.85 says that b is invertible in $\mathbf{B}$ if and only if $f_M(b) \neq 0$ for all $M \in \mathcal{M}(\mathbf{B})$, which is a repetition of Proposition 3.82.

In the other extreme case, a Banach algebra $\mathbf{B}$ with a trivial center, the only remaining choice is $\mathbf{A} = \mathbb{C}e$. But then $M = \{0\}$ is the only element of $\mathcal{M}(\mathbf{A})$, and consequently, $J_M = \{0\}$ as well, which leaves us with $\mathbf{B}_M = \mathbf{B}/\{0\} \cong \mathbf{B}$ and $b_M = b + \{0\}$ for all $b \in \mathbf{B}$. The statement of Proposition 3.85 is less breathtaking in that case: "b is invertible in $\mathbf{B}$ if and only if b is invertible in $\mathbf{B}$."

Remark 3.86. a) It should be mentioned that, for the application of Proposition 3.85, the algebra $\mathbf{A}$ does not have to be all of the center of $\mathbf{B}$. Hence, it is not even necessary to know $\operatorname{cen} \mathbf{B}$ to apply central localization. However, by the choice of $\mathbf{A}$, one can control the simplicity of the local algebras on a scale from $\mathbb{C}$ to $\mathbf{B}$, as the discussion of the two extreme cases above nicely illustrates. As a rule of thumb, the larger we choose $\mathbf{A}$, the larger are the maximal ideals M and the local ideals J_M, and the finer are the local algebras $\mathbf{B}_M$.

b) If one does not know the whole maximal ideal space of a commutative Banach algebra $\mathbf{B}$, one can still pass to a unital subalgebra $\mathbf{A}$ of $\mathbf{B}$, for which $\mathcal{M}(\mathbf{A})$ is known, and proceed as above instead of applying Gelfand theory. ☐

Allan-Douglas localization vs. limit operators

It was first pointed out in [67] that the limit operator concept is compatible with the local principle by ALLAN and DOUGLAS. We will restrict ourselves to saying a few words about that. For details see Section 4 of [67] or Section 2.3.5 of [70].

Let A be a band-dominated operator on $E = \ell^p(\mathbb{Z}^n, \mathbf{X})$. What we will do is to study the invertibility at infinity of A, that is invertibility in the non-commutative Banach algebra $\mathrm{BDO}^p/K(E, \mathcal{P})$ by means of Allan-Douglas localization without even thinking about our limit operator approach. But we will soon realize that what we automatically end up if we follow this road are limit operators.

To use the local principle for $\mathbf{B} = \mathrm{BDO}^p/K(E, \mathcal{P})$, we have to fix a Banach subalgebra $\mathbf{A}$ of the center of $\mathbf{B}$. For this subalgebra we will make the following choice. Put

$$\mathbf{A} := \{ \, \hat{M}_f + K(E, \mathcal{P}) \; : \; f \in C(\overline{\mathbb{R}^n}) \, \}.$$

It is readily seen that $\hat{M}_f$, with $f \in C(\overline{\mathbb{R}^n})$, commutes with the generators of BDO^p modulo $K(E, \mathcal{P})$ so that $\mathbf{A}$ is indeed contained in the center of $\mathbf{B}$. Moreover, $\mathcal{M}(\mathbf{A})$ can be identified with S^{n-1}, where we identify $s \in S^{n-1}$ with the character

$$\hat{M}_f + K(E, \mathcal{P}) \;\mapsto\; \lim_{x \to \infty_s} f(x)$$

of $\mathbf{A}$. Following the local principle, we let J_s denote the smallest ideal of $\mathbf{B}$ containing the maximal ideal of $\mathbf{A}$ that is associated with $s \in S^{n-1}$, and abbreviate the local algebra $\mathbf{B}/J_s = (\mathrm{BDO}^p/K(E, \mathcal{P}))/J_s$ by $\mathbf{B}_s$ and the local representative $A + K(E, \mathcal{P}) + J_s$ by A_s for every $s \in S^{n-1}$. Then Proposition 3.85 involves the invertibility of A_s in $\mathbf{B}_s$ for every $s \in S^{n-1}$. But regarding to this, we have the following very nice coincidence as shown in Theorem 6 of [67].

Proposition 3.87. *Let $A \in \mathrm{BDO}^p_{\$}$ and $s \in S^{n-1}$. Then A_s is invertible in $\mathbf{B}_s$ if and only if the local operator spectrum $\sigma^{\mathrm{op}}_s(A)$ is uniformly invertible.*

Involving the local principle, Proposition 3.85, we then get the following.

Corollary 3.88. *An operator $A \in \mathrm{BDO}^p_{\$}$ is invertible at infinity if and only if $\sigma^{\mathrm{op}}_s(A)$ is uniformly invertible for every $s \in S^{n-1}$.*

Note that this is an even stronger result than Theorem 1 since it does not require the uniform boundedness of the suprema

$$\sup_{B \in \sigma^{\mathrm{op}}_s(A)} \|B^{-1}\|$$

with respect to $s \in S^{n-1}$! Therefore, Corollary 3.88 was an important step towards getting rid of the uniform boundedness condition in Theorem 1 – as far as possible. Section 3.9 below is devoted to this question. Another step into this direction is shown in what follows.

Our choice of the central subalgebra $\mathbf{A}$ and the rather simple behaviour of a function $f \in C(\overline{\mathbb{R}^n})$ towards ∞_s for every $s \in S^{n-1}$ had the effect that every local representative $A_s \in \mathbf{B}_s$ represents the whole local operator spectrum $\sigma^{\mathrm{op}}_s(A)$ which can still contain quite a lot of operators. We could get a finer localization of $A + K(E, \mathcal{P})$ if we found a larger subalgebra of $\mathrm{cen}\,\mathbf{B}$.

Remember from Remark 3.53 that $b \in \ell^\infty(\mathbb{Z}^n, \mathbb{C}I)$ is slowly oscillating if and only if $V_\alpha b - b \in \ell^\infty_0$ for all $\alpha \in \mathbb{Z}^n$. But this is clearly equivalent to

$$[V_\alpha, \, \hat{M}_b] \;=\; (V_\alpha \hat{M}_b V_{-\alpha} - \hat{M}_b) V_\alpha \;=\; \hat{M}_{V_\alpha b - b} V_\alpha \;\in\; K(E, \mathcal{P})$$

for all $\alpha \in \mathbb{Z}^n$. Let us denote the set of all slowly oscillating $b \in \ell^\infty(\mathbb{Z}^n, \mathbb{C}I)$ by $\mathrm{SO}(\mathbb{Z}^n)$. Since $\hat{M}_b$ clearly commutes with every generalized multiplication operator if $b \in \mathrm{SO}(\mathbb{Z}^n)$, we get that

$$\mathbf{A} := \{ \hat{M}_b + K(E, \mathcal{P}) : b \in \mathrm{SO}(\mathbb{Z}^n) \} \tag{3.44}$$

is contained in the center of $\mathbf{B} = \mathrm{BDO}^p / K(E, \mathcal{P})$. In fact, it follows (see [61] or Theorem 2.4.2 of [70]) that (3.44) **coincides** with the center of $\mathrm{BDO}_{\$}^p / K(E, \mathcal{P})$.

This observation gives rise to a much finer localization of this factor algebra, culminating in an analogue of Proposition 3.87 where $S^{n-1} \cong \mathcal{M}_\infty(C(\overline{\mathbb{R}^n}))$ is replaced by $\mathcal{M}_\infty(\mathrm{SO}(\mathbb{Z}^n))$. Therefore, one has to appropriately redefine the local operator spectra $\sigma_s^{\mathrm{op}}(A)$ for $s \in \mathcal{M}_\infty(\mathrm{SO}(\mathbb{Z}^n))$ by using nets instead of sequences due to the topological nature of $\mathcal{M}(\mathrm{SO}(\mathbb{Z}^n))$. As a consequence, also the analogue of Corollary 3.88 holds where S^{n-1} is replaced by $\mathcal{M}_\infty(\mathrm{SO}(\mathbb{Z}^n))$. Note that this is another, even stronger, improvement of Theorem 1. As a corollary of this result we get Proposition 3.113 since in that case the localization is fine enough that $\sigma_s^{\mathrm{op}}(A)$ turns out to consist of a single operator only for every $s \in \mathcal{M}_\infty(\mathrm{SO}(\mathbb{Z}^n))$.

3.6 Generalizations of the Limit Operator Concept

There is a number of possible generalizations of the limit operator concept as introduced here. Subject to possible modifications in the definition

$$A_h := \mathcal{P}\text{-}\lim U_{h_m}^{-1} A U_{h_m}$$

are

- the family $\{U_r\}$ of invertible isometries,

- the approximate identity $\mathcal{P}$, and

- the space E,

where a certain compatibility between these three items is needed, of course. We will briefly present one sensible choice of the family $\{U_r\}$ for the case $E = L^p$, which differs from the setting presented in the rest of this book.

For every $r \in \mathbb{R}_+$, put

$$U_r := V_\vartheta \Phi_r^{-1},$$

where V_ϑ is our usual shift operator on L^p by a fixed vector $\vartheta \in \mathbb{R}^n$, and

$$(\Phi_r u)(x) = \rho_r \, u\left(\frac{x}{r}\right)$$

blows the argument of a function $u \in E$ up by the factor $r \in \mathbb{R}_+$ with

$$\rho_r = \begin{cases} r^{-\frac{n}{p}} & p < \infty, \\ 1 & p = \infty \end{cases}$$

being a correction factor that makes Φ_r an isometry on $E = L^p = L^p(\mathbb{R}^n)$.

With this construction, the isometry $U_r^{-1} = \Phi_r V_{-\vartheta}$ magnifies and shifts the local behaviour of a function $u \in E$ at the fixed point $\vartheta \in \mathbb{R}^n$ to the origin.

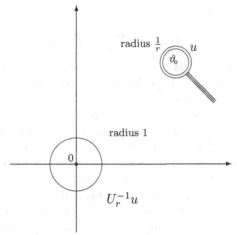

Figure 10: The behaviour of u in a ball of radius $1/r$ around ϑ
is magnified and shifted to the unit ball centered at the origin.

Now the operator

$$U_r^{-1}AU_r \;=\; \Phi_r V_{-\vartheta} A V_\vartheta \Phi_r^{-1}$$

(shrink, shift, apply A, shift back, blow up) studies the behaviour of A in a neighbourhood of the point ϑ, which is of interest, for example, if a lot of oscillation is
going on at ϑ. Limit operators can be defined here by letting the zoom ratio r go
to infinity in a prescribed way $r_1, r_2, \ldots$.

For an application of limit operators in this zooming context see [6], where the
local behaviour of a singular integral operator with slowly oscillating coefficients
at the endpoint ϑ of a Carleson curve in the plane is studied.

Of course, in this connection one has to replace our approximate identity
$\mathcal{P}$ by another one which is more appropriate for this process of zooming into ϑ
instead of shifting towards infinity.

3.7   Examples

It is time to look at some important examples of band-dominated operators and
their operator spectra.

3.7.1   Characteristic Functions of Half Spaces

Let $\langle \cdot, \cdot \rangle$ refer to the standard inner product in $\mathbb{R}^n$, fix a non-zero vector $a \in \mathbb{Z}^n$,
and denote by

$$U = \{x \in \mathbb{R}^n \;:\; \langle a, x \rangle \geq 0\} \tag{3.45}$$

the half space in $\mathbb{R}^n$ for which the boundary ∂U of U passes through the origin
and is orthogonal to the vector a, and a is contained in U. Then the multiplication

operator $P_U = M_{\chi_U}$ is a band-dominated operator on $E = L^p$ for every $p \in [1, \infty]$. Recall from basic geometry that $|\langle a, x \rangle| = |a|_2 \operatorname{dist}_2(x, \partial U)$ for every $x \in \mathbb{R}^n$.

Proposition 3.89. *The operator P_U is rich, and its operator spectrum consists of*

(i) *the identity operator I,*

(ii) *the zero operator 0, and*

(iii) *all shifts P_{c+U} of P_U with $c \in \mathbb{Z}^n$.*

Proof. Put $A = P_U$ and take an arbitrary sequence in $\mathbb{Z}^n$ tending to infinity. As in the proof of Proposition 3.9, choose a subsequence $h = (h_m)$ with $h_m \to \infty_s$ for some direction $s \in S^{n-1}$. Depending on $\langle a, s \rangle$, we have the following three cases.

(i) If $\langle a, s \rangle > 0$, then $\langle a, h_m \rangle \to +\infty$, which shows that $A_h = I$.

(ii) If $\langle a, s \rangle < 0$, then $\langle a, h_m \rangle \to -\infty$, which shows that $A_h = 0$.

(iii) If $\langle a, s \rangle = 0$, then $\langle a, h_m \rangle \in \mathbb{Z}$ either remains bounded for all m or not. In the first case, there is a subsequence $g = (g_m)$ of h such that $\langle a, g_m \rangle \in \mathbb{Z}$ is a constant sequence. But then $V_{-g_m} A V_{g_m} = P_{-g_m + U}$ is constant since $c + U = d + U$ if and only if $c - d \in \partial U$; that is $\langle a, c \rangle = \langle a, d \rangle$. Consequently, $A_g = P_{c+U}$ with a $c \in \mathbb{Z}^n$. In the second case, we can choose a subsequence $g = (g_m)$ of h such that $\langle a, g_m \rangle$ tends to plus or minus infinity, which brings us to either (i) or (ii).

Conversely, it is easily seen that every operator of the form P_{c+U} with $c \in \mathbb{Z}^n$ is in the operator spectrum of A, by choosing $h_m = mb - c$ for example, with a vector $b \in \partial U \cap \mathbb{Z}^n$. $\qquad\square$

As a corollary we get that $P_U = P_{U_1} \cdots P_{U_k}$ is rich if $U = U_1 \cap \cdots \cap U_k$ is a polyhedron with half spaces $U_1, \ldots, U_k \subset \mathbb{R}^n$. We will briefly discuss the simple case $k = 2$ in $\mathbb{R}^2$. Therefore, let U_1 and U_2 denote two half planes in $\mathbb{R}^2$ of the form (3.45) such that $\partial U_1 \neq \partial U_2$, and put $U = U_1 \cap U_2$.

Corollary 3.90. *The operator P_U is rich, and its operator spectrum consists of*

(i) *the identity operator I,*

(ii) *the zero operator 0,*

(iii) *all shifts P_{c+U_1} of P_{U_1} with $c \in \mathbb{Z}^n$, and*

(iv) *all shifts P_{c+U_2} of P_{U_2} with $c \in \mathbb{Z}^n$.*

Proof. From $A := P_U = P_{U_1} P_{U_2}$ and Propositions 3.11 a) and 3.4 c) we get that A is rich and that every limit operator of A is the product of two limit operators (towards the same direction) of P_{U_1} and P_{U_2}, which covers the cases (i)–(iv) by Proposition 3.89. Conversely, as in the proof of Proposition 3.89, for every operator (i)–(iv), we can easily find a sequence $h \subset \mathbb{Z}^n$ leading to the corresponding limit operator of A. $\qquad\square$

We will later apply this result to the quarter planes $U = U_1 \cap U_2$ and $V = V_1 \cap V_2$ with $U_1 = \{(x,y) \; : \; x \geq 0\}$, $U_2 = \{(x,y) \; : \; y \geq 0\}$, $V_1 = \{(x,y) \; : \; y \geq x\}$ and $V_2 = \{(x,y) \; : \; y \leq -x\}$.

3.7.2   Wiener-Hopf and Toeplitz Operators

Let A be a shift invariant operator on $E = L^p(\mathbb{R}^2)$, and put $U = U_1 \cap U_2$ with two half spaces U_1 and U_2 of the form (3.45). We will study the compression $P_U A P_U$ of A. At one point, we extend the operator $P_U A P_U$ by the complementary projector $Q_U = I - P_U$ in order to give it a chance to be invertible at infinity.

Lemma 3.91. *Let $A \in \mathrm{BDO}_S^p$ be shift invariant. Then the following holds.*

a) *A is invertible at infinity if and only if A is invertible.*

b) *If $P_U A P_U + Q_U$ is invertible at infinity, then A is invertible.*

c) *$\|P_U A P_U\| = \|A\|$.*

Proof. The "if" part is trivial (see Remark 2.5), and the "only if" part follows from Proposition 3.12 and $A \in \sigma^{\mathrm{op}}(A) = \{A\}$.

b) This also follows from Proposition 3.12 and $A \in \sigma^{\mathrm{op}}(P_U A P_U + Q_U)$.

c) Clearly, $\|P_U A P_U\| \leq \|P_U\|\|A\|\|P_U\| = \|A\|$. On the other hand, we have $\|A\| \leq \|P_U A P_U\|$ by Proposition 3.4 a) since $A \in \sigma^{\mathrm{op}}(P_U A P_U)$. $\square$

Since A is band-dominated and shift invariant, we get that $P_U A P_U$ is band-dominated, rich, and its operator spectrum consist of A, 0 and compressions of A to all integer shifts of U_1 and U_2 by Corollary 3.90.

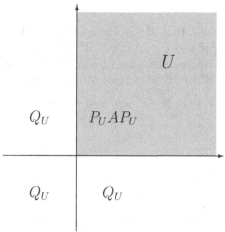

Figure 11: The Wiener-Hopf operator $P_U A P_U$ is extended to the whole space by adding the complementary projector Q_U of P_U. In this simple example, U is the intersection of the upper and the right half plane U_1 and U_2 in $\mathbb{R}^2$.

Proposition 3.92. *The operator $P_U A P_U + Q_U$ is invertible at infinity if and only if $P_{U_1} A P_{U_1}$ and $P_{U_2} A P_{U_2}$ are invertible on $L^p(U_1)$ and $L^p(U_2)$, respectively.*

Proof. From Theorem 1 we know that $P_U A P_U + Q_U$ is invertible at infinity if and only if its limit operators A, I,

$$P_{c+U_1} A P_{c+U_1} + Q_{c+U_1} \quad \text{and} \quad P_{c+U_2} A P_{c+U_2} + Q_{c+U_2} \tag{3.46}$$

are invertible for all $c \in \mathbb{Z}^2$ and their inverses are uniformly bounded. This clearly proves the "only if" part.

For the "if" part, suppose the two half plane compressions of A to U_1 and U_2 are invertible on $L^p(U_1)$ and $L^p(U_2)$, respectively. Then their extensions

$$A_1 = P_{U_1} A P_{U_1} + Q_{U_1} \quad \text{and} \quad A_2 = P_{U_2} A P_{U_2} + Q_{U_2}$$

and all shifts of these are invertible on L^p. But since A is shift invariant, the shifts $V_c A_1 V_{-c}$ and $V_c A_2 V_{-c}$ coincide with the operators (3.46) above. Finally, the invertibility of A follows from Lemma 3.91 b), and the inverses of all these limit operators of $P_U A P_U + Q_U$ are bounded by $\max\{1, \|A^{-1}\|, \|A_1^{-1}\|, \|A_2^{-1}\|\}$.   □

We will now look at some shift invariant operators A on L^p and on ℓ^p.

If A is a convolution operator[7] on $L^p(\mathbb{R}^2)$, then the compression $P_U A P_U$ is referred to as a two-dimensional *Wiener-Hopf operator* [87]. If the convolution kernel of A is in $L^1(\mathbb{R}^2)$, then A is band-dominated (even $A \in \mathcal{W}$), and we have Proposition 3.92 to study whether or not this Wiener-Hopf operator is invertible at infinity.

The discrete analogue of a convolution operator on L^p is a band-dominated operator on ℓ^p with constant coefficients. Such operators are also known as *Laurent operators*[8]. The compression of a Laurent operator A to a proper subset U of $\mathbb{Z}^n$ is typically referred to as a *discrete Wiener-Hopf operator* or a *Toeplitz operator*.

Note that the results of Subsection 3.7.1 literally translate to the discrete case ℓ^p if we replace the half spaces (3.45) by their intersections with $\mathbb{Z}^n$. Consequently, also Proposition 3.92 applies to the discrete case with the replacement of half spaces by discrete half spaces.

To be able to draw a matrix, we pass to $n = 1$, and hence to one-dimensional Laurent operators as in Example 1.39. Put $U = \mathbb{N}_0$ and

$$T(a) = P_U L(a) P_U,$$

where $a \in L^\infty(\mathbb{T})$ and $L(a)$ is defined as in Example 1.39. The one-dimensional Toeplitz operator $T(a)$ is typically studied on $\ell^p(\mathbb{N}_0)$ but we will embed it into $\ell^p(\mathbb{Z})$ by studying $T(a) + Q_{\mathbb{N}_0}$ in order to use our techniques.

Since 0 and I are the only limit operators of P_U, and since $L(a)$ is shift invariant, $T(a) + Q_U$ has only $L(a)$ and I as its limit operators. This is nicely

---

[7]This is the analogue of Example 1.45 in two dimensions. Also see Subsection 4.2.1.
[8]Remember Example 1.39 for the one-dimensional case $n = 1$.

confirmed by a look at the following matrix.

$$[T(a) + Q_U] = \begin{pmatrix} \ddots & & & 0 & & & & & & 0 \\ & 1 & & & & & & & & \\ & & 1 & & & & & & & \\ 0 & & & 1 & & & & & & \\ \hline & & & & a_0 & a_{-1} & a_{-2} & \ddots & & \\ & & & & a_1 & a_0 & a_{-1} & a_{-2} & \ddots & \\ & & & & a_2 & a_1 & a_0 & a_{-1} & \ddots & \\ & & & & \ddots & a_2 & a_1 & a_0 & \ddots & \\ 0 & & & & & \ddots & \ddots & \ddots & \ddots & \end{pmatrix} \qquad (3.47)$$

If we travel towards $+\infty$; that is down the diagonal, we arrive at $L(a)$, and towards $-\infty$; that is up the diagonal, we arrive at I.

From Theorem 1 we consequently get that $T(a) + Q_U$ is invertible at infinity if and only if $L(a)$ is invertible.

3.7.3   Singular Integral Operators

By evaluating the following integral in the sense of the Cauchy principal value, the singular integral operator

$$(Su)(t) = \frac{1}{\pi i} \int_{\mathbb{T}} \frac{u(t)}{s-t}\, ds, \qquad t \in \mathbb{T}$$

is well-defined and bounded on a dense subspace of $L^2(\mathbb{T})$, whence it can be extended to a bounded linear operator on $L^2(\mathbb{T})$. It is readily seen that, via the Fourier isomorphism between $L^2(\mathbb{T})$ and $\ell^2 = \ell^2(\mathbb{Z})$, the singular integral operator S corresponds to the operator

$$S_{\ell^2} = \hat{M}_{\mathrm{sgn}} = P - Q \qquad (3.48)$$

with

$$\mathrm{sgn}(m) = \begin{cases} 1, & m \geq 0, \\ -1, & m < 0, \end{cases} \qquad m \in \mathbb{Z},$$

$P = P_{\mathbb{N}_0}$ and $Q = I - P$.

Recall from Example 1.39 that the operator M_c of multiplication by a function $c \in L^\infty(\mathbb{T})$ on $L^2(\mathbb{T})$ corresponds to the Laurent operator $L(c)$ on ℓ^2 via the Fourier isomorphism. Now suppose c, d are continuous functions on $\mathbb{T}$, and we will study the bounded linear operator

$$A = M_c + M_d S \qquad (3.49)$$

on $L^2(\mathbb{T})$. From (3.48) we get that A can be identified with the operator

$$
\begin{aligned}
A_{\ell^2} &= L(c) + L(d) S_{\ell^2} = L(c) (P+Q) + L(d) (P-Q) \\
&= L(c+d) P + L(c-d) Q = L(a) P + L(b) Q
\end{aligned}
\tag{3.50}
$$

on ℓ^2, where $a = c + d$, $b = c - d \in C(\mathbb{T})$.

From Example 1.39 we know that A_{ℓ^2} is band-dominated, and

$$
[A_{\ell^2}] =
\begin{pmatrix}
\ddots & \ddots & \ddots & & & & & \\
\ddots & b_0 & b_1 & \ddots & & & & \\
\ddots & b_{-1} & b_0 & b_1 & \ddots & & & \\
& \ddots & b_{-1} & b_0 & a_1 & \ddots & & \\
& & \ddots & b_{-1} & a_0 & a_1 & \ddots & \\
& & & & a_{-1} & a_0 & a_1 & \ddots \\
& & & & \ddots & a_{-1} & a_0 & \ddots \\
& & & & & \ddots & \ddots & \ddots
\end{pmatrix}
.
\tag{3.51}
$$

The obvious limit operators of P, Q, $L(a)$ and $L(b)$ or just a look at this matrix immediately show that $\sigma^{\mathrm{op}}(A_{\ell^2}) = \{L(a), L(b)\}$, with $L(a)$ in the local operator spectrum at $+\infty$ and $L(b)$ at $-\infty$. So A_{ℓ^2}, and consequently A, is Fredholm[9] if and only if $L(a)$ and $L(b)$ are invertible. With the same argument for $A - \lambda I$ and another look back to Example 1.39, we get that

$$
\mathrm{sp_{ess}} \, A = \mathrm{sp_{ess}} \, A_{\ell^2} = a(\mathbb{T}) \cup b(\mathbb{T})
$$

since $A_{\ell^2} - \lambda I = L(a-\lambda)P + L(b-\lambda)Q$ is Fredholm if and only if $a - \lambda$ and $b - \lambda$ are invertible in $L^\infty(\mathbb{T})$. Moreover, by Corollary 3.19, the Fredholm index of A is equal to $\mathrm{ind}^+ L(a) + \mathrm{ind}^- L(b)$ which, by basic results on Toeplitz operators [10], is the difference of the winding numbers of b and a with respect to the origin.

If $b \in L^\infty(\mathbb{T})$ is invertible, we can restrict ourselves to the study of $L(a)P+Q$ instead of A_{ℓ^2}, where we denote the function $b^{-1}a$ by a again. For this operator we have

$$
L(a)P + Q = \Big(PL(a)P + Q\Big)\Big(I + QL(a)P\Big) = \Big(T(a) + Q\Big)\Big(I + QL(a)P\Big)
$$

where the second factor $I + QL(a)P$ is always an invertible operator with its inverse equal to $I - QL(a)P$. This equality shows that, in this case, the study of the operator (3.49) is equivalent to the study of the Toeplitz operator (3.47) from Subsection 3.7.2.

---

[9]Recall Figure 4 on page 57 and Theorem 1.

3.7.4 Discrete Schrödinger Operators

In this subsection we discuss the discrete Schrödinger operator

$$A = \sum_{k=1}^{n}(V_{e_k} + V_{-e_k}) + \hat{M}_c,$$

on ℓ^p where $e_1, \ldots, e_n$ are the unit vectors in $\mathbb{R}^n$, and $c \in \ell^\infty(\mathbb{Z}^n, L(\mathbf{X}))$ is called the potential of A. This operator is the discrete analogue of the second order differential operator $-\Delta + M_b$ where Δ is the Laplacian and $b \in L^\infty$ is a bounded potential, both on $\mathbb{R}^n$. A is a band operator with band-width 1. For $n = 1$, its matrix representation is

$$[A] = \begin{pmatrix} \ddots & \ddots & & & & & & 0 \\ \ddots & c_{-2} & 1 & & & & & \\ & 1 & c_{-1} & 1 & & & & \\ & & 1 & c_0 & 1 & & & \\ & & & 1 & c_1 & 1 & & \\ & & & & 1 & c_2 & \ddots & \\ 0 & & & & & \ddots & \ddots \end{pmatrix}.$$

Let $1 < p < \infty$ and $\dim \mathbf{X} < \infty$. Then $\mathbf{X}$ is reflexive and $A \in \mathrm{BO} \subset \mathcal{W}$ is rich by Corollary 3.24. In this case, A is Fredholm if and only if it is invertible at infinity, which, on the other hand, is equivalent to the elementwise invertibility of $\sigma^{\mathrm{op}}(A)$ by Proposition 3.112. This argument with $A - \lambda I$ in place of A shows that we can improve (3.12) to

$$\mathrm{sp}_{\mathrm{ess}}\, A = \bigcup_{B \in \sigma^{\mathrm{op}}(A)} \mathrm{sp}\, B. \tag{3.52}$$

Depending on the function c, one can give a more or less explicit description of the operator spectrum $\sigma^{\mathrm{op}}(A)$ as well as of $\mathrm{sp}\, B$ for $B \in \sigma^{\mathrm{op}}(A)$, and hence, of the essential spectrum of A.

- If c is almost periodic, then the $\mathcal{P}$-convergence in Definition 3.3 is automatically convergence in the norm of $L(E)$, and the invertibility of a limit operator of A immediately implies that of A. Repeating this argument with $A - \lambda I$ in place of A, we get that $\mathrm{sp}\, B = \mathrm{sp}\, A$ for all $B \in \sigma^{\mathrm{op}}(A)$, whence (3.52) translates to

$$\mathrm{sp}_{\mathrm{ess}}\, A = \mathrm{sp} A.$$

In particular, A is Fredholm if and only if all $B \in \sigma^{\mathrm{op}}(A)$ are invertible, which is the case if and only if A is invertible.

- If c is slowly oscillating, then $\sigma^{\mathrm{op}}(\hat{M}_c) = \{aI : a \in c(\infty)\}$. But then all limit operators of A are Laurent operators. For these operators the spectrum is well known (see [9], for example). Together with (3.52), this leads to

$$\mathrm{sp}_{\mathrm{ess}}\, A = \bigcup_{B \in \sigma^{\mathrm{op}}(A)} \mathrm{sp}\, B = c(\infty) + [-2N, 2N],$$

which is Theorem 2.6.25 in [70].

- Finally, if c is pseudo-ergodic, then $\sigma^{\mathrm{op}}(\hat{M}_c)$ is the set of all operators $\hat{M}_d$ with a function $d : \mathbb{Z}^n \to \mathrm{clos}\, c(\mathbb{Z})$, and

$$\sigma^{\mathrm{op}}(A) = \left\{ \sum_{k=1}^{n}(V_{e_k} + V_{-e_k}) + \hat{M}_d \quad : \quad d : \mathbb{Z}^n \to \mathrm{clos}\, c(\mathbb{Z}) \right\}.$$

Consequently, A itself is among its own limit operators, which shows that A is Fredholm if and only if it is invertible, and

$$\mathrm{sp}_{\mathrm{ess}}\, A = \bigcup_{B \in \sigma^{\mathrm{op}}(A)} \mathrm{sp}\, B = \mathrm{sp}\, A.$$

For the limit operator method in the treatment of the essential spectrum of discrete Schrödinger operators for some more classes of potentials, see [64].

3.8 Limit Operators – Everything Simple Out There?

In the introduction to this chapter we motivated the limit operator method as a tool to study the invertibility at infinity of A via localization "over infinity",

$$A \mapsto \sigma^{\mathrm{op}}(A) = \{A_h\}. \tag{3.53}$$

We also mentioned that the main idea of localization is to substitute the original problem by a family of simpler invertibility problems. Of course, the image of a homomorphism (remember Subsection 3.5.2) is an algebra with a simpler structure, and hence the invertibility problem there cannot be harder than the original one. But the question that naturally comes up in connection with the map (3.53) is whether, and in which sense, the limit operators A_h are simpler objects than the operator A itself.

If we think of generalized or usual multiplication operators A with a slowly oscillating symbol, then the limit operators of A are multiples of the identity, which are certainly very simple objects. On the other hand we have the results on generalized multiplication operators A with almost periodic or pseudo-ergodic symbol. In the first case, the invertibility of any limit operator of A implies the invertibility of A, which shows that every single limit operator represents quite a lot of the behaviour of A. In the second case, A itself is even among its limit operators.

Here are a view quick-and-dirty arguments, ordered from true to true but shallow, why A_h is a simpler – or at least not more complicated – object than A:

- $A \in \mathrm{BDO}_{\mathcal{S}}^p$ being invertible at infinity is sufficient for A_h to be invertible.

- Every submatrix of $[A_h]$ is (approximately equal to) a submatrix of $[A]$.

- A "sample" of A cannot possess more complexity than A itself.

To find out a bit more about the aspect of simplification when passing from A to A_h, we will address each of the following two questions in a little subsection.

- Is this step of simplification ultimate in the sense that things are not going to become simpler when doing this step again?

- Is it a "privilege" for an operator to be limit operator of another one? Do limit operators form some subclass of "simple" operators?

3.8.1   Limit Operators of Limit Operators of ...

We have shown that some interesting properties (including invertibility at infinity, Fredholmness and applicability of approximation methods) of A are closely connected with all limit operators of A (or an associated operator) being invertible. But if verifying the latter property is easier than looking at A itself, then perhaps things become even easier when we pass to limit operators again. And again and again...?

In this subsection we will show that every limit operator of a limit operator of A is already a limit operator of A. We will do this in two steps:

- Show that $V_{-\alpha} B V_\alpha \in \sigma^{\mathrm{op}}(A)$ if $\alpha \in \mathbb{Z}^n$ and $B \in \sigma^{\mathrm{op}}(A)$.

- Show that the $\mathcal{P}$-limit of a sequence in $\sigma^{\mathrm{op}}(A)$ is in $\sigma^{\mathrm{op}}(A)$.

To give us a bit more fluent language, we will add one more definition.

Definition 3.93. *Fix an arbitrary operator $A \in L(E)$. Operators of the form $V_{-\alpha} A V_\alpha$ with $\alpha \in \mathbb{Z}^n$ are called shifts of A, and the set of all shifts of A will be denoted by $\mathcal{V}_A$,*

$$\mathcal{V}_A := \{ V_{-\alpha} A V_\alpha \}_{\alpha \in \mathbb{Z}^n}.$$

We will generalize this notation to sets $L \subset L(E)$ of operators by putting

$$\mathcal{V}_L := \bigcup_{A \in L} \mathcal{V}_A = \{ V_{-\alpha} A V_\alpha \ : \ \alpha \in \mathbb{Z}^n, A \in L \}.$$

So, by what we defined earlier, $A \in L(E)$ is a shift invariant operator if and only if $\mathcal{V}_A = \{A\}$. We will moreover call a set $L \subset L(E)$ *shift invariant* if $\mathcal{V}_L = L$.

Proposition 3.94. *For $A \in L(E)$, the operator spectrum $\sigma^{op}(A)$ as well as all local operator spectra $\sigma_s^{op}(A)$ with $s \in S^{n-1}$ are shift invariant sets.*

Proof. Take $A \in L(E)$ and let $h = (h_m) \subset \mathbb{Z}^n$ be a sequence tending to infinity such that $A_h \in \sigma^{op}(A)$ exists. Now take an arbitrary $\alpha \in \mathbb{Z}^n$ and verify that the sequence $h + \alpha := (h_m + \alpha) \subset \mathbb{Z}^n$ leads to

$$A_{h+\alpha} = \mathcal{P}\text{-}\lim_{m\to\infty} V_{-\alpha} V_{-h_m} A V_{h_m} V_\alpha = V_{-\alpha} (\mathcal{P}\text{-}\lim_{m\to\infty} V_{-h_m} A V_{h_m}) V_\alpha = V_{-\alpha} A_h V_\alpha.$$

Clearly, $h + \alpha$ tends to infinity in the same direction as h does. $\qquad\square$

One task down, still one to be done:

Proposition 3.95. *For $A \in L(E)$, the global operator spectrum as well as all local operator spectra are sequentially closed with respect to $\mathcal{P}$-convergence.*

Proof. We will give the proof for the global operator spectrum. It is completely analogous for local operator spectra.

Given a sequence $A^{(1)}, A^{(2)}, \ldots \subset \sigma^{op}(A)$ which $\mathcal{P}$-converges to $B \in L(E)$, we will show that also $B \in \sigma^{op}(A)$. For $k = 1, 2, \ldots$, let $h^{(k)} = (h_m^{(k)}) \subset \mathbb{Z}^n$ denote a sequence tending to infinity such that $A^{(k)} = A_{h^{(k)}}$, and choose a subsequence $g^{(k)}$ out of $h^{(k)}$ such that, for all $m \in \mathbb{N}$, the following holds

$$\left\| P_m \left(V_{-g_m^{(k)}} A V_{g_m^{(k)}} - A^{(k)} \right) \right\| < \frac{1}{m}.$$

Now we put $g = (g_m) := (g_m^{(m)})$ and show that $B = A_g$. Therefore, take a $r \in \mathbb{N}$ and some $\varepsilon > 0$. Then, for all $m > r$,

$$\begin{aligned}
\|P_r(V_{-g_m} A V_{g_m} - B)\| &\leq \|P_r P_m (V_{-g_m} A V_{g_m} - A^{(m)})\| + \|P_r(A^{(m)} - B)\| \\
&\leq \|P_r\| \, \|P_m(V_{-g_m^{(m)}} A V_{g_m^{(m)}} - A^{(m)})\| \\
&\qquad\qquad\qquad\qquad + \|P_r(A^{(m)} - B)\| \\
&\leq 1 \cdot 1/m \quad + \quad \|P_r(A^{(m)} - B)\|
\end{aligned}$$

holds, which clearly tends to zero as $m \to \infty$. The same can be shown for P_r coming from the right, and so we have $B = A_g \in \sigma^{op}(A)$. $\qquad\square$

Corollary 3.96. *For $A \in L(E)$, the global operator spectrum as well as all local operator spectra are closed subsets of $L(E)$.*

Proof. Every norm-convergent sequence in $\sigma^{op}(A)$ is $\mathcal{P}$-convergent of course, and, by Proposition 3.95, its limit is in $\sigma^{op}(A)$ again. $\qquad\square$

If we take Propositions 3.94 and 3.95 together, we get that the operator spectrum is closed under shifting and under passing to $\mathcal{P}$-limits. The same is true for local operator spectra. Consequently, we have:

Corollary 3.97. *Every limit operator of a limit operator of $A \in L(E)$ is already a limit operator of A itself.*

So no further operators – and hence no further simplification – occurs when repeatedly passing to the set of limit operators.

3.8.2 Everyone is Just a Limit Operator!

Another question which is closely related with the mentioned effect of simplification is whether or not the set of all possible limit operators

$$\bigcup_{A \in L(E, \mathcal{P})} \sigma^{op}(A)$$

is some proper subset of $L(E, \mathcal{P})$ containing the "simple" operators only. Here the answer is "No". The main result of this subsection essentially says that no operator is "complicated" enough not to be the limit operator of an even more "complicated" one.

So take some $p \in [1, \infty]$ and a Banach space $\mathbf{X}$, put $E = \ell^p(\mathbb{Z}^n, \mathbf{X})$, and take an arbitrary operator $A \in L(E, \mathcal{P})$. The following construction of an "even more complicated operator than A" was suggested by ROCH. The basic idea is very similar to the explicit construction of a sequence s with a prescribed set $\{a_1, a_2, \ldots\}$ of partial limits by putting $s := (\, a_1\, ,\, a_1, a_2\, ,\, a_1, a_2, a_3\, ,\, \ldots\,)$.

We still need some preparations. Firstly, choose and fix a sequence $h = (h_m) \subset \mathbb{Z}^n$ that tends to infinity fast enough that all sets

$$U_m := h_m + \{-m, m\}^n, \qquad m = 1, 2, \ldots \tag{3.54}$$

are pairwise disjoint (for instance, put $h_m = ((m+1)^2, \ldots, (m+1)^2)$).

Given a bounded sequence $(A_m)_{m=1}^\infty$ with $A_m \in L(\operatorname{im} P_m)$ for every $m \in \mathbb{N}$, we let $S(A_1, A_2, \ldots)$ refer to the operator on E for which every space $\operatorname{im} P_{U_m}$ is an invariant subspace of E, and S acts on $\operatorname{im} P_{U_m}$ as $V_{h_m} A_m V_{-h_m}$, i.e. as the operator A_m, shifted to $\operatorname{im} P_{U_m}$. Otherwise, $S(A_1, A_2, \ldots)$ be the zero operator, that is

$$\big(S(A_1, A_2, \ldots)u\big)_\alpha = \begin{cases} (V_{h_m} A_m V_{-h_m} u)_\alpha & \text{if } \alpha \in U_m, \ m \in \mathbb{N}, \\ 0 & \text{otherwise} \end{cases} \tag{3.55}$$

for every $u \in E$. In this sense, one can also write

$$S(A_1, A_2, \ldots) = \sum_{m=1}^\infty V_{h_m} A_m V_{-h_m},$$

the sum in $(Su)_\alpha$ actually being finite for every $\alpha \in \mathbb{Z}^n$ by the choice of $h = (h_m)$.

Since $S(A_1, A_2, \ldots)$ acts as a block-diagonal operator, we immediately have that $S(A_1, A_2, \ldots) \in L(E)$ with

$$\|S(A_1, A_2, \ldots)\| = \sup_{m \in \mathbb{N}} \|A_m\|. \tag{3.56}$$

From (3.55) we easily get that, for every $\alpha \in \mathbb{Z}^n$, $P_\alpha S(A_1, A_2, \ldots)Q_m \rightrightarrows 0$ and $Q_m S(A_1, A_2, \ldots)P_\alpha \rightrightarrows 0$ as $k \to \infty$ which clearly implies $S(A_1, A_2, \ldots) \in L(E, \mathcal{P})$.

Lemma 3.98. *The map* $(A_m) \mapsto S(A_1, A_2, \ldots)$ *is bounded and linear from the space of all bounded sequences* $(A_m)_{m=1}^\infty$ *with* $A_m \in L(\operatorname{im} P_m)$ *to* $L(E, \mathcal{P})$.

Proof. The equalities

$$S(A_1 + B_1, A_2 + B_2, \ldots) = S(A_1, A_2, \ldots) + S(B_1, B_2, \ldots)$$

and

$$S(\lambda A_1, \lambda A_2, \ldots) = \lambda \, S(A_1, A_2, \ldots)$$

for all such sequences (A_m) and (B_m) and all $\lambda \in \mathbb{C}$ follow immediately from the definition (3.55), and the boundedness follows from (3.56). $\qquad\square$

Now remember our operator $A \in L(E, \mathcal{P})$, put $A_m := P_m A P_m$ for every $m \in \mathbb{N}$, and define

$$S_A := S(P_1 A P_1, P_2 A P_2, \ldots).$$

Then we can say even more about $\|S_A\|$ than in (3.56). Since $A \in L(E, \mathcal{P})$, we get from Proposition 1.70 that $A_m \xrightarrow{\mathcal{P}} A$ and that $\|A\| \leq \sup \|A_m\|$. Together with (3.56) and $\|A_m\| = \|P_m A P_m\| \leq \|A\|$, this shows that

$$\|S_A\| = \sup_{m \in \mathbb{N}} \|A_m\| = \|A\|. \tag{3.57}$$

Lemma 3.99. *If A is a band operator, then S_A is a band operator of the same band-width, and if A is band-dominated, then so is S_A.*

Proof. If A is a band operator, then also $V_{h_m} A_m V_{-h_m}$ is one with the same band-width, and hence also S_A has this property. If A is band-dominated, then we can take a sequence of band operators with $A^{(k)} \rightrightarrows A$. Now also the $S_{A^{(k)}}$ are band operators, and Lemma 3.98 and (3.57) show that

$$\|S_{A^{(k)}} - S_A\| = \|S_{A^{(k)} - A}\| = \|A^{(k)} - A\| \to 0$$

as $k \to \infty$, showing that also S_A is band-dominated. $\qquad\square$

Now we are ready for the main result of this subsection, saying that every $A \in L(E, \mathcal{P})$ is a limit operator – even with respect to a prescribed direction towards infinity.

Proposition 3.100. *For every $A \in L(E, \mathcal{P})$ and every sequence $g = (g_m) \in \mathbb{Z}^n$ tending to infinity, there is an operator $B \in L(E, \mathcal{P})$ and a subsequence h of g such that $A = B_h$.*

Proof. From the sequence g choose a subsequence $h = (h_m)$ such that the sets U_m, defined by (3.54), are pairwise disjoint, and put $B = S_A$. From the definition of S_A we know that, for every $m \in \mathbb{N}$,

$$P_{U_m} B = V_{h_m} P_m A P_m V_{-h_m} = B P_{U_m} \tag{3.58}$$

and consequently

$$P_m V_{-h_m} B V_{h_m} = V_{-h_m} P_{U_m} B V_{h_m} = P_m A P_m$$

holds. Now fix $k \in \mathbb{N}$. Then, for every $m > k$,

$$P_k V_{-h_m} B V_{h_m} = P_k P_m V_{-h_m} B V_{h_m} = P_k P_m A P_m = P_k A P_m,$$

and hence

$$P_k(A - V_{-h_m} B V_{h_m}) = P_k A - P_k A P_m = P_k A Q_m \rightrightarrows 0$$

as $m \to \infty$ since $A \in L(E, \mathcal{P})$. The symmetric property $(A - V_{-h_m} B V_{h_m}) P_k \rightrightarrows 0$ as $m \to \infty$ is analogously derived from the second equality in (3.58). This shows that $V_{-h_m} B V_{h_m} \overset{\mathcal{P}}{\to} A$ as $m \to \infty$ and completes the proof.   □

Remark 3.101. It was shown that $\sigma^{\mathrm{op}}(S_A)$ contains A. By the same idea, putting

$$S_{A,B} := S(\, P_1 A P_1 \, , \, P_2 B P_2 \, , \, P_3 A P_3 \, , \, P_4 B P_4 \, , \, \ldots \,),$$

one can construct an operator which has the two operators A and B in its operator spectrum. This idea can be extended up to finding an operator S with countably many prescribed limit operators in the same local operator spectrum!   □

3.9   Big Question: Uniformly or Elementwise?

It is the biggest question of the whole limit operator business whether or not in Theorem 1, and consequently in all its applications, **uniform** invertibility of $\sigma^{\mathrm{op}}(A)$ can be replaced by **elementwise** invertibility. In other words:

> **Big question:**   *Is the operator spectrum of a rich operator automatically uniformly invertible if it is elementwise invertible?*

The fact that no one has found a counter-example yet, of course doesn't mean that there is no such example. On the other hand, the whole theory would immediately gain a lot of simplification and beauty if one were able to give a positive answer to the big question in general.

Our strategy to tackle this problem is as follows. Sets with the word "spectrum" in their name surprisingly often enjoy some sort of compactness property. If the operator spectrum $\sigma^{\mathrm{op}}(A)$ were a compact subset of $L(E)$, we could immediately answer the big question.

Indeed, suppose the answer were "No". Then we had a rich operator $A \in L(E)$ whose operator spectrum is elementwise but not uniformly invertible. Choose a sequence $A_1, A_2, \ldots$ from $\sigma^{op}(A)$ such that $\|A_m^{-1}\| \to \infty$ as $m \to \infty$. From the compactness of $\sigma^{op}(A)$ we conclude that (A_m) has a subsequence with norm-limit B in $\sigma^{op}(A)$ again. But since B is invertible, by our premise, Lemma 1.2 b) contradicts the fact of the growing inverses of A_m. So the answer had to be "Yes".

Unfortunately, the operator spectrum $\sigma^{op}(A)$ in general is not compact with respect to the norm topology in $L(E)$. But we will follow this road a little bit further and discover that $\sigma^{op}(A)$, however, enjoys a weaker form of compactness.

The content of this section should be seen as an appetizer to the big question of limit operator business and to some of the problems and strategies to tackle it, finally leading to a positive answer for $E = \ell^p(\mathbb{Z}^n, \mathbf{X})$ in the cases $p = 1$ and $p = \infty$. Maybe this section can stimulate someone to tackle the problem for the remaining cases $p \in (1, \infty)$, where it seems to be most promising to start with the C^* case ℓ^2. Also note that the uniformity condition in Theorem 1 was already considerably weakened in [67], [61] and [70], using central localization techniques – remember Subsection 3.5.2 of this book or look at Sections 2.3 and 2.4 of [70].

3.9.1 Reformulating Richness

We start our investigations with a slight but very natural reformulation of an operator's rich-property. As usual, we will say that a set $L \subset L(E)$ is *relatively $\mathcal{P}$-sequentially compact* if every infinite sequence from L has a $\mathcal{P}$-convergent subsequence, and we will say that L is *$\mathcal{P}$-sequentially compact* if this $\mathcal{P}$-limit is in L again.

Proposition 3.102. *An operator $A \in L(E)$ is rich if and only if its set of shifts*

$$\mathcal{V}_A = \{V_{-\alpha} A V_\alpha\}_{\alpha \in \mathbb{Z}^n}$$

is relatively $\mathcal{P}$-sequentially compact.

Proof. If $\mathcal{V}_A$ is relatively $\mathcal{P}$-sequentially compact, then all infinite sequences $(V_{-\alpha_m} A V_{\alpha_m})_{m=1}^\infty$, including those with α_m tending to infinity as $m \to \infty$, possess a $\mathcal{P}$-convergent subsequence. So A is rich.

If conversely, A is rich, then every sequence $(V_{-\alpha_m} A V_{\alpha_m})$ with $\alpha_m \to \infty$ has a $\mathcal{P}$-convergent subsequence. All other sequences $(V_{-\alpha_m} A V_{\alpha_m}) \subset \mathcal{V}_A$ have an infinite subsequence such that all α_k involved are contained in some bounded set $U \subset \mathbb{Z}^n$. Since U has only finitely many elements, there is even an infinite constant subsequence of $(V_{-\alpha_m} A V_{\alpha_m})$ which, clearly, $\mathcal{P}$-converges. Summarizing this, every sequence $(V_{-\alpha_m} A V_{\alpha_m})$ in $\mathcal{V}_A$ (with α_k tending to infinity or not) has a $\mathcal{P}$-convergent subsequence. $\qquad\square$

Remark 3.103. In Subsection 3.4.13 we gave some arguments why we restrict ourselves to integer shift distances when approaching limit operators. Note that another nice fact that we would lose by passing to real valued shift distances is the only-if part of Proposition 3.102.

For instance, the operator $A := P_{[0,\infty)}$ on $L^p(\mathbb{R})$ is rich, but in $\{V_{-\alpha}AV_\alpha\}_{\alpha\in\mathbb{R}}$ there are many sequences, choose $\alpha_m := [m\sqrt{2}]$ for example, without a $\mathcal{P}$-convergent subsequence. In this connection we may also recall the operator of multiplication by the function f in Example 3.63 with rational step length s.   □

As trivial Proposition 3.102 seems, it opens the door to seeing rich operators and their operator spectra in a somewhat different light. For instance, we just have to recall Proposition 3.95 to find and prove the following interesting fact.

Proposition 3.104. *For a rich operator $A \in L(E)$, every local as well as the global operator spectrum of A is $\mathcal{P}$-sequentially compact.*

Proof. If $A \in L(E)$ is rich, then, by Proposition 3.102, $\mathcal{V}_A$ is relatively $\mathcal{P}$-sequentially compact. Consequently, $\mathcal{P}$-clos$\mathcal{V}_A$ is $\mathcal{P}$-sequentially compact. Now clearly, $\sigma^{op}(A)$, as well as every local operator spectrum of A, is a subset of the latter set. But as $\mathcal{P}$-sequentially closed subsets (cf. Proposition 3.95) of a $\mathcal{P}$-sequentially compact set, $\sigma^{op}(A)$ and $\sigma_s^{op}(A)$ are $\mathcal{P}$-sequentially compact themselves.   □

So once more, the word "spectrum" has turned out to be a good indicator for a compactness property.

Remark 3.105. Although this is not true for Proposition 3.102 (see Remark 3.103), Proposition 3.104 remains valid with some small changes in its proof when we define limit operators by using real shift distances.   □

3.9.2   Turning Back to the Big Question

The discovery that operator spectra of rich operators are $\mathcal{P}$-sequentially compact encourages us to think about the big question with a bit more optimism. It is a first clue towards believing that the answer might be "Yes" – the more beautiful but less expected one of the two possible answers.

However, for ordinary operators the answer has to be "No", as the following example shows.

Example 3.106. Let $(a_m)_{m=1}^\infty \subset [0,1)$ be a sequence of pairwise distinct numbers, and put $U_m := a_m + [0,1]$ for every $m \in \mathbb{N}$, so that all these intervals U_m differ from each other. Now put

$$A_m := \frac{1}{m}P_{U_m} + Q_{U_m}, \qquad m = 1, 2, \ldots$$

and let A denote the operator S from Remark 3.101, having all operators $A_1, A_2, \ldots$ in its operator spectrum. It turns out that the operator spectrum of A, besides $A_1, A_2, \ldots$ (and integer shifts of those), only contains the identity operator I, which is due to the odd choice of the intervals U_m. This set is elementwise invertible – but not uniformly since $\|A_m^{-1}\| = m$.

If this operator A were rich, then, by Proposition 3.104, the sequence (A_m) would have some $\mathcal{P}$-convergent subsequence which clearly is not possible due to the incompatibility of the intervals U_m. □

1st Try

So suppose $A \in L(E)$ is rich, and recall the situation from the beginning of this section; that is, $\sigma^{\text{op}}(A)$ is elementwise but not uniformly invertible. As before, choose a sequence $A_1, A_2, \dots$ from $\sigma^{\text{op}}(A)$ such that $\|A_m^{-1}\| \to \infty$ as $m \to \infty$.

This time, indeed, Proposition 3.104 says that (A_m) has a $\mathcal{P}$-convergent subsequence with $\mathcal{P}$-limit B in $\sigma^{\text{op}}(A)$. The question jumping to our mind now is whether this operator B can be invertible or not.

Answer: Although the norm-limit of a sequence, whose inverses tend to infinity, cannot be invertible, the $\mathcal{P}$-limit B can:

Example 3.107. The sequence

$$A_m := P_m + \frac{1}{m}Q_m, \qquad m = 1, 2, \dots \tag{3.59}$$

$\mathcal{P}$-converges to the identity I as $m \to \infty$, although its inverses are growing unboundedly with $\|A_m^{-1}\| = m$. □

Learning from this Setback

The reason why the sequence (3.59), despite of its badly growing inverses, however has an invertible $\mathcal{P}$-limit, is that, roughly speaking, those "parts" of A_m which are responsible for the growing inverses, are "running away" towards infinity when $m \to \infty$, and so they have no contribution to the $\mathcal{P}$-limit B.

Luckily, we here have to do with operator spectra and those have one more nice feature, stated in Proposition 3.94: We stay in $\sigma^{\text{op}}(A)$ if we pass to appropriate shifts $V_{-\alpha_m} A_m V_{\alpha_m}$ instead of A_m, where every $\alpha_m \in \mathbb{Z}^n$ shall be chosen such that the "bad parts" of $V_{-\alpha_m} A_m V_{\alpha_m}$ cannot run away to infinity any more and consequently, they must have some "bad impact" on the $\mathcal{P}$-limit B as well!

That is the idea. In order to write it down in a more precise way, we extend the notion from Definition 2.31. Recall that an operator $A \in L(E)$ is said to be bounded below if

$$\nu(A) := \inf_{\|x\|=1} \|Ax\| > 0.$$

We will moreover say that a set $L \subset L(E)$ is *uniformly bounded below* if every $A \in L$ is bounded below and if there is a $\nu > 0$ such that $\nu(A) \geq \nu$ for all $A \in L$; that is

$$\|Ax\| \geq \nu\|x\|, \qquad A \in L, \ x \in E.$$

Now take a Banach space $\mathbf{X}$ and put $E = \ell^p(\mathbb{Z}^n, \mathbf{X})$. The following proposition gives a positive answer to the big question for operators in $\text{BDO}_\$^p$ with $p = \infty$. In fact, it is an even stronger statement than what we need.

Proposition 3.108. *If $p = \infty$, $A \in \mathrm{BDO}^p_{\$}$ and all limit operators of A are bounded below, then $\sigma^{\mathrm{op}}(A)$ is uniformly bounded below.*

Proof. Let $p = \infty$, $A \in \mathrm{BDO}^p_{\$}$, and suppose $\sigma^{\mathrm{op}}(A)$ is elementwise but not uniformly bounded below. Then there is a sequence $(A_m) \subset \sigma^{\mathrm{op}}(A)$ such that, for every $m \in \mathbb{N}$, there is an $x_m \in E$ with

$$\|x_m\| = 1 \qquad \text{and} \qquad \|A_m x_m\| \leq \frac{1}{m}.$$

Now fix some non-empty bounded set $U \subset \mathbb{Z}^n$. For our purposes, meaning $p = \infty$, we may in fact put $U := \{0\}$. Clearly, there are translation vectors $c_m \in \mathbb{Z}^n$ such that

$$\|P_{c_m + U} x_m\|_\infty > \|x_m\|_\infty / 2 = 1/2$$

for every[10] $m \in \mathbb{N}$. Now put

$$y_m := V_{-c_m} x_m \qquad \text{and} \qquad B_m := V_{-c_m} A_m V_{c_m}, \tag{3.60}$$

which leaves us with

$$\|y_m\| \quad = \quad \|V_{-c_m} x_m\| = \|x_m\| = 1, \tag{3.61}$$

$$\|P_U y_m\| \quad = \quad \|P_U V_{-c_m} x_m\| = \|P_{c_m + U} x_m\| > \frac{1}{2} \|y_m\| \tag{3.62}$$

and

$$\|B_m y_m\| \quad = \quad \|V_{-c_m} A_m V_{c_m} V_{-c_m} x_m\| = \|A_m x_m\| \leq \frac{1}{m} \tag{3.63}$$

for every $m \in \mathbb{N}$. Moreover, by Proposition 3.94, the sequence (B_m) is in $\sigma^{\mathrm{op}}(A)$ again. So, by Proposition 3.104, we can pass to a subsequence of (B_m) with $\mathcal{P}$-limit $B \in \sigma^{\mathrm{op}}(A)$. For simplicity of notations, suppose (B_m) itself already be this sequence.

Now some of the "bad parts" of B_m are located inside U, and hence, they cannot "run away" when $m \to \infty$. In fact, now we can prove that the $\mathcal{P}$-limit B has inherited some "bad parts" inside U – bad enough to make B not bounded below:

As an element of $\sigma^{\mathrm{op}}(A)$, the operator B is bounded below by our premise. So there is a constant $a := \nu(B) > 0$ such that

$$\|Bx\| \geq a\|x\| \qquad \text{for all} \qquad x \in E. \tag{3.64}$$

Now fix some continuous function $\varphi : \mathbb{R}^n \to [0, 1]$ which is identically 1 on $H = [0, 1)^n$ and vanishes outside of $2H$. By φ_r denote the function

$$\varphi_r(t) := \varphi(t/r), \qquad t \in \mathbb{R}^n$$

---

[10] And this is a typical ℓ^∞-argument! ℓ^p is much more sophisticated here since U has to be chosen sufficiently large, and it is not clear if the same U works for all x_m.

for every $r > 0$, so that φ_r is supported in $2rH$, while on rH it is equal to 1.

From $A \in \mathrm{BDO}_\$^p \subset \mathrm{BDO}^p$, Proposition 3.6 b), Theorem 1.42 and $\varphi \in \mathrm{BUC}$ we know that $[B, \hat{M}_{\varphi_r}] \rightrightarrows 0$ as $r \to \infty$. Choose $r \in \mathbb{N}$ large enough that $U + H \subset rH$ and

$$\| [B, \hat{M}_{\varphi_r}] \| < \frac{a}{6}. \tag{3.65}$$

Since $B_m \xrightarrow{P} B$, there is some $m_0 \in \mathbb{N}$ such that

$$\| P_{2r}(B - B_m) \| < \frac{a}{6} \qquad \text{for all} \qquad m > m_0. \tag{3.66}$$

From $\hat{M}_{\varphi_r} = \hat{M}_{\varphi_r} P_{2rH}$ and inequalities (3.65) and (3.66) we conclude that

$$
\begin{aligned}
\| B\hat{M}_{\varphi_r} - \hat{M}_{\varphi_r} B_m \| &\le \| B\hat{M}_{\varphi_r} - \hat{M}_{\varphi_r} B \| + \| \hat{M}_{\varphi_r}(B - B_m) \| \\
&\le \| B\hat{M}_{\varphi_r} - \hat{M}_{\varphi_r} B \| + \| \hat{M}_{\varphi_r} \| \cdot \| P_{2rH}(B - B_m) \| \\
&< \frac{a}{6} + 1 \cdot \frac{a}{6} = \frac{a}{3} \qquad \text{for all} \qquad m > m_0.
\end{aligned}
$$

The latter shows that, for every $x \in E$,

$$
\begin{aligned}
\| B\hat{M}_{\varphi_r} x \| &\le \| \hat{M}_{\varphi_r} B_m x \| + \| B\hat{M}_{\varphi_r} x - \hat{M}_{\varphi_r} B_m x \| \\
&\le \| \hat{M}_{\varphi_r} B_m x \| + \| B\hat{M}_{\varphi_r} - \hat{M}_{\varphi_r} B_m \| \cdot \| x \| \\
&\le \| \hat{M}_{\varphi_r} B_m x \| + \frac{a}{3} \| x \|
\end{aligned}
\tag{3.67}
$$

holds if $m > m_0$. Taking everything together, keeping in mind that $P_U = P_U \hat{M}_{\varphi_r}$ since $U + H \subset rH$ and φ_r is equal to 1 on rH, we get

$$
\begin{aligned}
\frac{a}{2} &\overset{(3.61)}{=} \frac{a}{2} \| y_m \| \overset{(3.62)}{<} a \| P_U y_m \| = a \| P_U \hat{M}_{\varphi_r} y_m \| \le a \| P_U \| \cdot \| \hat{M}_{\varphi_r} y_m \| \\
&= a \| \hat{M}_{\varphi_r} y_m \| \overset{(3.64)}{\le} \| B\hat{M}_{\varphi_r} y_m \| \overset{(3.67)}{\le} \| \hat{M}_{\varphi_r} B_m y_m \| + \frac{a}{3} \| y_m \| \\
&\overset{(3.61)}{\le} \| \hat{M}_{\varphi_r} \| \cdot \| B_m y_m \| + \frac{a}{3} \cdot 1 \overset{(3.63)}{\le} 1 \cdot \frac{1}{m} + \frac{a}{3}
\end{aligned}
$$

for all $m > m_0$. If we finally subtract $a/3$ at both ends of the chain, we arrive at

$$\frac{a}{6} \le \frac{1}{m} \qquad \text{for all} \qquad m > m_0,$$

which, of course, is contradicting $a > 0$. $\qquad \square$

This result implies a positive answer to the big question for $p = 1$ and $p = \infty$:

Theorem 3.109. *If $p \in \{1, \infty\}$, then the operator spectrum of a rich band-dominated operator is uniformly invertible as soon as it is elementwise invertible.*

Proof. Let $p = \infty$, and suppose $A \in \mathrm{BDO}_\$^\infty$ has an elementwise invertible operator spectrum $\sigma^{\mathrm{op}}(A)$. Then, by Lemma 2.35, all limit operators of A are bounded below, and Proposition 3.108 shows that they are even uniformly bounded below. But then, by Lemma 2.35 again, the norms of their inverses are bounded from above, which proves the theorem for $p = \infty$.

Now let $p = 1$, and suppose $A \in \mathrm{BDO}_\$^p$. Then $A^* \in \mathrm{BDO}_\$^\infty$, and the operator spectra $\sigma^{\mathrm{op}}(A)$ and $\sigma^{\mathrm{op}}(A^*)$ can be identified elementwise by taking (pre-)adjoints. Since this identification clearly preserves norms and invertibility, we are done with $p = 1$ as well. $\square$

Well, that was quite something. However, speaking with Sir Edmund Hillary, the big "bastard" still to be "knocked off", is the big question for the case $p \in (1, \infty)$, which is a bit ironic since this tended to be the nice and handsome case before, as we remember from Figures 1, 3 and 4 (see pages 15, 46 and 57), for example.

3.9.3   Alternative Proofs for ℓ^∞

After the proof presented in the previous subsection was found in 2003, it turned out that a positive answer to the big question in the case $\ell^\infty = \ell^\infty(\mathbb{Z}^n)$ was already hidden in the middle of the proof of Theorem 8 in [67]. Although this is a theorem on the Wiener algebra $\mathcal{W}$, a closer look shows that this particular step works for all band-dominated operators on ℓ^∞.

Meanwhile, a third proof for $E = \ell^\infty$ has been found by CHANDLER-WILDE and the author in [17]. One of the first messages of this paper is that the notions of convergence studied by CHANDLER-WILDE and ZHANG in [24] are compatible with many of the concepts presented in this book. For example:

- The strict convergence $u_m \xrightarrow{\beta} u$ in E of [24] is equivalent to $u_m \xrightarrow{\mathcal{P}} u$ in E.

- The class of operators $A \in L(E)$ which map strictly convergent to strictly convergent sequences in E is the superset of $L(E, \mathcal{P})$ of operators $A \in L(E)$ which are subject to the first constraint of (1.12).

- The class of operators $A \in L(E)$ which map strictly convergent to norm convergent sequences in E is the superset of $K(E, \mathcal{P})$ of operators $A \in L(E)$ which are subject to the constraint (1.10).

Then [17] goes on to apply the collective compactness approach of [24] to the operator spectrum of a rich band-dominated operator. Therefore, we briefly introduce the following notions of [24]:

A set $L \subset L(E)$ is called *collectively sequentially compact on* (E, s) if, for every bounded set $B \subset E$, the image of B under the set L is relatively sequentially compact in E with respect to strict convergence.

For a sequence $(A_m) \subset L(E)$ and an operator $A \in L(E)$, one writes $A_m \overset{\beta}{\to} A$ if $u_m \overset{\beta}{\to} u$ implies $A_m u_m \overset{\beta}{\to} Au$. A subset $K \subset L$ is called β-dense in $L \subset L(E)$ if, for every $A \in L$, there is a sequence $(A_m) \subset K$ with $A_m \overset{\beta}{\to} A$.

Finally, a set $\mathcal{V}$ of isometrical isomorphisms on E is called *sufficient* if, for some $m \in \mathbb{N}$, it holds that, for every $u \in E$, there exists $V \in \mathcal{V}$ such that $2 \|P_m V u\| \geq \|u\|$. Then Theorem 4.5 in [24] says the following.

Proposition 3.110. *Suppose that $\mathcal{V}$ is a sufficient set of isometrical isomorphisms on E, and $L \subset L(E)$ has the following properties:*

(i) *L is collectively sequentially compact on (E, s);*

(ii) *L is β-sequentially compact;*

(iii) *$V^{-1}BV \in L$ for all $B \in L$ and $V \in \mathcal{V}$;*

(iv) *$I - B$ is injective for all $B \in L$.*

Then the set $I - L$ is uniformly bounded below. If, moreover, in $I - L$ there is an s-dense subset of surjective operators, then all operators in $I - L$ are surjective.

This result can be applied to

$$E = \ell^\infty, \qquad \mathcal{V} = \{V_\alpha\}_{\alpha \in \mathbb{Z}^n} \qquad \text{and} \qquad L = \sigma^{\mathrm{op}}(I - A)$$

with a rich band-dominated operator A on E. Indeed:

(o) $\mathcal{V}$ is sufficient if $E = \ell^\infty$ since every $u \in \ell^\infty$ can be shifted such that at least 50% of its norm are attained in the 0th component.

(i) Since $\mathcal{P} \subset K(E)$ if $\dim \mathbf{X} < \infty$, a set $L \subset L(E)$ is collectively sequentially compact on (E, s) if and only if L is bounded. The latter is true for the operator spectrum by Proposition 3.4 a).

(ii) It is readily shown that $A_m \overset{P}{\to} A$ implies $A_m \overset{\beta}{\to} A$. Therefore, and by Proposition 3.104, we know that the operator spectrum of a rich operator is β-sequentially compact.

(iii) From Proposition 3.94 we get that $V^{-1}BV$ is a limit operator of $I - A$ if B is one and if $V \in \mathcal{V}$.

Now Proposition 3.110 yields the result of Proposition 3.108 and hence the desired answer to the big question for ℓ^∞.

3.9.4 Passing to Subclasses

Although, apart from $p \in \{1, \infty\}$, we are not yet able to definitely answer the big question in general, there are some subclasses of $\mathrm{BDO}_\$^p$ for which a positive answer can be given, independently of p. If the set $A + \mathbb{C}I$ is contained in one of these classes, then, for $\dim \mathbf{X} < \infty$, we have the opposite inclusion to (3.12)

$$\bigcup_{B \in \sigma^{\mathrm{op}}(A)} \mathrm{sp}\, B \supset \mathrm{sp}_{\mathrm{ess}}\, A, \qquad (3.68)$$

which shows, together with (3.12), that

$$\bigcup_{B \in \sigma^{\mathrm{op}}(A)} \mathrm{sp}\, B = \mathrm{sp}_{\mathrm{ess}}\, A, \qquad (3.69)$$

if we are in the so-called perfect case, $1 < p < \infty$ and $\dim \mathbf{X} < \infty$, and for the set $A + \mathbb{C}I$ the elementwise invertibility of operator spectra implies their uniform invertibility.

A first but very simple situation in which elementwise invertibility implies uniform invertibility of $\sigma^{\mathrm{op}}(A)$ is when $\sigma^{\mathrm{op}}(A)$ is a compact subset of $L(E)$, as we already found out earlier. For example, this is the case when $\sigma^{\mathrm{op}}(A)$ is contained in a finite-dimensional subspace of $L(E)$ since, by Proposition 3.4 a) and Corollary 3.96, the operator spectrum is always a bounded and closed subset of $L(E)$.

Some more sophisticated subclasses with a positive answer to the big question are studied in Sections 2.3 – 2.5 of [70]. Here we will basically cite two of the highlights of this theory and refer to [70] for the proofs.

Operators in the Wiener Algebra

Fix a Banach space $\mathbf{X}$, and put $E = \ell^p(\mathbb{Z}^n, \mathbf{X})$ where $p \in [1, \infty]$ may vary if we say so. The definition of a limit operator of $A \in L(E)$ is based on the notion of $\mathcal{P}$-convergence which, in general, depends on the norm of the underlying space E and hence on the value of $p \in [1, \infty]$. So even if $A \in \mathrm{BDO}^p$ for several values of p, the operator spectrum $\sigma^{\mathrm{op}}(A)$ can differ from p to p. But for operators A in the Wiener algebra

$$\mathcal{W} \subset \mathrm{BDO}^p \qquad \text{for all} \qquad p \in [1, \infty],$$

(recall Definition 1.43) one can show that $\sigma^{\mathrm{op}}(A)$ does not depend on p.

Lemma 3.111. *If $A \in \mathcal{W}$ is rich and $h \subset \mathbb{Z}^n$ tends to infinity, then there is a subsequence g of h such that the limit operator A_g exists with respect to all spaces E with $p \in [1, \infty]$. This limit operator again belongs to $\mathcal{W}$, and $\|A_g\|_{\mathcal{W}} \leq \|A\|_{\mathcal{W}}$.*

Proof. This is Proposition 2.5.6 in [70]. $\square$

As a consequence of Proposition 1.46 c) we get that, if $A \in \mathcal{W}$ is invertible on one space E with $p \in [1, \infty]$, then it is invertible on all these spaces E, and $\|A^{-1}\|_{L(E)} \leq \|A^{-1}\|_{\mathcal{W}}$. Moreover, a consequence of Lemma 3.111 is that the operator spectrum $\sigma^{\mathrm{op}}(A)$ is contained in the Wiener algebra $\mathcal{W}$ and does not depend on the underlying space E if $A \in \mathcal{W}$. So for $A \in \mathcal{W}$, the statement of Theorem 1 holds independently of the underlying space E. Moreover, by Theorem 3.109, also the uniform boundedness condition of the inverses is redundant since this is true for $p \in \{1, \infty\}$ and consequently also for $p \in (1, \infty)$ by Riesz-Thorin interpolation (see [39, IV.1.4] or [84, V.1]). So, for operators in the Wiener algebra, we are in the following nice situation.

Proposition 3.112. *If* $\mathbf{X}$ *is reflexive,* $E = \ell^p(\mathbb{Z}^n, \mathbf{X})$ *with some* $p \in [1, \infty]$, *and* $A \in W$ *is rich, then the following statements are equivalent.*

(i) *A is invertible at infinity on one of the spaces E.*

(ii) *All limit operators of A are invertible on one of the spaces E.*

(iii) *A is invertible at infinity on all spaces E.*

(iv) *All limit operators of A are invertible on all spaces E.*

Proof. This is Theorem 2.5.7 of [70]. $\qquad\square$

So, in particular, the big question has a positive answer for every operator $A \in W$, which includes all band operators.

Band-dominated Operators with Slowly Oscillating Coefficients

Again fix a Banach space $\mathbf{X}$ and some $p \in [1, \infty]$, and put $E = \ell^p(\mathbb{Z}^n, \mathbf{X})$. Remember from Remark 3.53 that a sequence $b \in \ell^\infty(\mathbb{Z}^n, L(\mathbf{X}))$ is called slowly oscillating if, for every $q \in \mathbb{Z}^n$, the sequence $V_q b - b$ vanishes at ∞.

From the discrete analogue of Corollary 3.48 we know, as already formulated in Remark 3.53, that if A is a band-dominated operator on E with slowly oscillating coefficients, then A is rich and all limit operators of A have constant coefficients; that is, they are shift-invariant.

In addition to this simple structure of each limit operator of A, we have the following result on the operator spectrum of A which says that the elementwise invertibility of $\sigma^{\mathrm{op}}(A)$ implies its uniform invertibility.

Proposition 3.113. *If* $A \in \mathrm{BDO}_{\mathcal{S}}^p$ *has slowly oscillating coefficients, then A is invertible at infinity if and only if all limit operators of A are invertible.*

Proof. This is Theorem 2.4.27 in [70]. $\qquad\square$

So also in this case, the uniform boundedness of the inverses turns out to be redundant.

3.10 Comments and References

What is probably the first application of limit operators goes back to 1927, where FAVARD [28] used them to study ordinary differential equations with almost-periodic coefficients. Since that time limit operators have been used in the context of partial differential and pseudo-differential operators and in many other fields of numerical analysis (for instance, see [35] and [36]). The Fredholmness of pseudo-differential operators in spaces of functions on $\mathbb{R}^n$ has been studied by MUHAMADIEV in [54], [55], and by LANGE and RABINOVICH in [45] and [59, 60]. The first time limit operator techniques were applied to the general case of band-dominated operators was in 1985 by LANGE and RABINOVICH [43, 44]. After the break-through [67], RABINOVICH, ROCH and SILBERMANN and a small number of their co-authors pushed the research in this field forward to the current standard which is manifested in the monography [70], the current 'bible' of the limit operator method.

The results in Section 3.1 are standard (see [67], [68], etc.). Part b) of Proposition 3.11 goes back to [50]. The history of Theorem 1 starts with [67] for the 'perfect case' $1 < p < \infty$ with $\dim \mathbf{X} < \infty$ and [68] for L^2. The cases $\dim \mathbf{X} = \infty$ and $p \in \{1, \infty\}$ were then covered by [70], [47] and [50]. Lemma 3.14 basically goes back to SIMONENKO [83].

The index results in Subsection 3.3.1 go back to [66] via computations of the K-group of the C^*-algebra BDO^2 for $p = 2$, and [74] for the generalization to $1 < p < \infty$. It should be mentioned that the results in [74] easily extend to operators in BDO^1 and BDO^∞ that belong to the Wiener algebra $\mathcal{W}$. Subsection 3.3.2 is from [17]. Note that even stronger results are presented in [63] and in Section 6.3 of [70] for the C^*-algebra case; that is when $p = 2$ and $\mathbf{X}$ is a Hilbert space. Much more can be said about the approximation of spectra and pseudo-spectra for particular classes of band-dominated operators, for example, see [5] for banded Toeplitz operators (compressions of band operators with constant coefficients, see Subsection 3.7.2, in particular (3.47)).

Section 3.4 is an extended version of [49]. Operators with pseudo-ergodic coefficients have been studied by DAVIES in [26]. The compatibility of the limit operator method and central localization, as indicated in Subsection 3.5.2, was first shown in [67]. The observation that the center of $\mathrm{BDO}^p_\$/K(E, \mathcal{P})$ coincides with (3.44) and the localization of $\mathrm{BDO}^p_\$/K(E, \mathcal{P})$ over $\mathcal{M}_\infty(\mathrm{SO}(\mathbb{Z}^n))$ goes back to [61]. Generalizations of the limit operator approach, as indicated in Section 3.6, can be found in [62] and Chapter 7 of [70]. Proposition 3.95 is essentially shown in [67] already. Corollary 3.97 is Corollary 1.2.4 in [70]. The results of Subsection 3.8.2 can be found in Subsection 2.1.5 of [70].

The 'big question', that is discussed in Section 3.9, is as old as Theorem 1. In his review of the article [61], ALBRECHT BÖTTCHER writes about the uniform boundedness condition in Theorem 1: "Condition $(*)$ is nasty to work with.". We have got nothing to add to this. Propositions 3.102, 3.104 and 3.108 go back to the author [50]. Subsection 3.9.3 is due to CHANDLER-WILDE and the author [17], and the results of Subsection 3.9.4 are from [67] and [61].

Chapter 4

Stability of the Finite Section Method

4.1 The Finite Section Method: Stability Versus Limit Operators

The two preceding chapters suggest an approach to the applicability question of an approximation method in terms of limit operators since the following three conditions are equivalent for a fairly large class of methods (A_τ).

- (A) The approximation method (A_τ) is stable.
- (B) The stacked operator $\oplus A_\tau$ is invertible at infinity.
- (C) The operator spectrum of $\oplus A_\tau$ is uniformly invertible.

We will demonstrate this approach when (A_τ) is the finite section method for operators on the axis, where we are able to derive pretty explicit criteria on the applicability of the method. So in what follows, n equals 1, and the operator sequence (A_τ) is

$$(A_{\lceil \tau \rceil}) = (P_\tau A P_\tau + Q_\tau), \qquad \tau \in T, \tag{4.1}$$

where $T = \mathbb{N}$ or $\mathbb{R}_+$. Note that this is the method (2.13) with $\Omega_\tau = \{-\tau, \ldots, \tau\}$ or $[-\tau, \tau]$, respectively, depending on whether we are in the discrete or continuous case.

4.1.1 Limit Operators of Stacked Operators

For the study of the stacked operator $\oplus A_\tau$, as introduced in Definition 2.18, we have to jump back and forth between operators on the axis and operators on the plane. Moreover, for reasons of analogy, we will study the discrete and the continuous case at once. To establish this polymorphism, we will reuse the notations of Section 2.4.

Throughout the following, let $p \in [1, \infty]$, and let E denote either the space $L^p = L^p(\mathbb{R})$ or $\ell^p(\mathbb{Z}, \mathbf{X})$ where $\mathbf{X}$ is an arbitrary Banach space. Depending on the choice of E; that is, whether we are in the continuous case or in the discrete case, take $\mathbb{D} \in \{\mathbb{R}, \mathbb{Z}\}$ corresponding, and define $E' = L^p(\mathbb{R}^2)$ or $E' = \ell^p(\mathbb{Z}^2, \mathbf{X})$, respectively. To make a distinction between operators on E and operators on E', we will write I', P'_U, P'_k, Q'_U, Q'_k for the identity operator and the respective projection operators on E'. Finally, put $\mathbb{D}_+ = \{\tau \in \mathbb{D} : \tau > 0\}$; that is $\mathbb{D}_+ \in \{\mathbb{R}_+, \mathbb{N}\}$, in accordance with the choice of $\mathbb{D}$.

Now suppose A is a rich band-dominated operator on E, with a pre-adjoint in case $p = \infty$; that is $A \in \mathrm{BDO}^p_{S,\$}$. We will show that $\oplus A_\tau$, stacked over the index set $T = \mathbb{D}_+$, as an operator on E', has all these properties as well, and we will give a description of its operator spectrum $\sigma^{\mathrm{op}}(\oplus A_\tau)$. The following formula, which is immediate from (4.1), will turn out to be helpful:

$$\oplus A_\tau = P'_V(\oplus A)P'_V + Q'_V, \tag{4.2}$$

where the quarter plane $V \subset \mathbb{D}^2$ is the intersection of the two half planes

$$H_{\mathsf{NE}} = \{(x, \tau) \in \mathbb{D}^2 : \tau \geq -x\} \quad \text{and} \quad H_{\mathsf{NW}} = \{(x, \tau) \in \mathbb{D}^2 : \tau \geq x\}.$$

Note that in (4.2) we may suppose that $\oplus A$ is stacked over $T = \mathbb{D}$. Moreover, we abbreviate

$$P'_{\mathsf{NE}} := P'_{H_{\mathsf{NE}}}, \quad Q'_{\mathsf{NE}} := I' - P'_{\mathsf{NE}}, \quad P'_{\mathsf{NW}} := P'_{H_{\mathsf{NW}}} \quad \text{and} \quad Q'_{\mathsf{NW}} := I' - P'_{\mathsf{NW}},$$

and put

$$\sigma^{\mathrm{op}}_M(B) := \bigcup_{\varphi \in M} \sigma^{\mathrm{op}}_{(\cos\varphi, \sin\varphi)}(B)$$

for all $B \in L(E')$ and all $M \in \{\mathsf{N}, \mathsf{S}, \mathsf{NE}, \mathsf{NW}\}$ where

$$\mathsf{N} = \left(\frac{\pi}{4}, \frac{3}{4}\pi\right), \quad \mathsf{S} = \left(\frac{3}{4}\pi, \frac{9}{4}\pi\right), \quad \mathsf{NE} = \left\{\frac{\pi}{4}\right\} \quad \text{and} \quad \mathsf{NW} = \left\{\frac{3}{4}\pi\right\}.$$

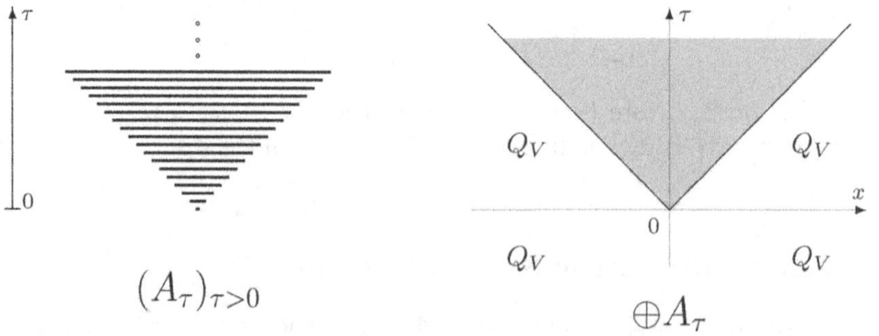

$$(A_\tau)_{\tau > 0} \qquad\qquad\qquad \oplus A_\tau$$

Figure 12: The finite section sequence (A_τ) and its stacked operator $\oplus A_\tau$ for $n = 1$.

Now here is the promised result.

Proposition 4.1. *If $A \in \mathrm{BDO}^p_{\mathcal{S},\$}$ on E, then $\oplus A_\tau \in \mathrm{BDO}^p_{\mathcal{S},\$}$ on E', and the operator spectrum $\sigma^{\mathrm{op}}(\oplus A_\tau)$ consists of the following operators:*

(i) *the identity operator I,*

(ii) *all operators $\oplus(V_{-c}AV_c)$ with $c \in \mathbb{Z}$,*

(iii) *all operators $\oplus A_h$ with $A_h \in \sigma^{\mathrm{op}}(A)$,*

and all shifts of

(iv) *the operators $P'_{\mathrm{NW}}(\oplus A_h) P'_{\mathrm{NW}} + Q'_{\mathrm{NW}}$ with $A_h \in \sigma^{\mathrm{op}}_+(A)$, and*

(v) *the operators $P'_{\mathrm{NE}}(\oplus A_h) P'_{\mathrm{NE}} + Q'_{\mathrm{NE}}$ with $A_h \in \sigma^{\mathrm{op}}_-(A)$,*

where all the $\oplus$ in (ii)–(v) are stacked over the index set $\mathbb{D}$.

Proof. Let $A \in \mathrm{BDO}^p_{\mathcal{S},\$}$. To show that $\oplus A_\tau \in \mathrm{BDO}^p_{\mathcal{S},\$}$, it is sufficient, by equation (4.2) and Proposition 3.11 a), to show that $\oplus A$, P'_V and hence $Q'_V = I' - P'_V$ are in $\mathrm{BDO}^p_{\mathcal{S},\$}$.

Clearly, the (generalized) multiplication operator P'_V is band-dominated and has a pre-adjoint. Moreover, by Corollary 3.90, it is rich, and we know everything about its operator spectrum. So it remains to check $\oplus A$. From Proposition 2.22 b) we get that $\oplus A \in \mathrm{BDO}^p$. Moreover, $(\oplus A)^\triangleleft = \oplus(A^\triangleleft)$ shows that $\oplus A \in \mathcal{S}$.

To see that $\oplus A$ is rich and to find out about its operator spectrum, we take an arbitrary sequence $h = (h_m)$ with $h_m = (h_m^{(1)}, h_m^{(2)}) \in \mathbb{Z}^2$ tending to infinity. Clearly, the second component $h^{(2)}$ of h has no influence at all on the operator $V_{-h_m}(\oplus A)V_{h_m}$. So we just have to look at the first component of h.

- If $h^{(1)}$ is bounded, then there is an infinite subsequence g of h and a $c \in \mathbb{Z}$ such that $g^{(1)}$ is the constant sequence c, and, consequently, $(\oplus A)_g$ exists and is equal to $\oplus(V_{-c}AV_c)$.

- If $h^{(1)}$ is unbounded, then h has a subsequence g the first component of which tends to plus or minus infinity. Since A is rich, we can choose a subsequence f from g such that $f^{(1)}$ leads to a limit operator of A. Hence $(\oplus A)_f$ exists and is equal to $\oplus(A_{f^{(1)}})$.

From Corollary 3.90, equation (4.1) and from what we just found out about the limit operators of $\oplus A$, we immediately get the inclusion "$\subset$" in the four equalities

$$\sigma^{\mathrm{op}}_{\mathrm{S}}(\oplus A_\tau) = \{I\},$$

$$\sigma^{\mathrm{op}}_{\mathrm{N}}(\oplus A_\tau) = \left\{ \oplus(V_{-c}AV_c) , \oplus A_h : c \in \mathbb{Z}, A_h \in \sigma^{\mathrm{op}}(A) \right\},$$

$$\sigma^{\mathrm{op}}_{\mathrm{NE}}(\oplus A_\tau) = \left\{ P'_U(\oplus A_h) P'_U + Q'_U : U = c + H_{\mathrm{NW}}, c \in \mathbb{Z}^2, A_h \in \sigma^{\mathrm{op}}_+(A) \right\},$$

$$\sigma^{\mathrm{op}}_{\mathrm{NW}}(\oplus A_\tau) = \left\{ P'_U(\oplus A_h) P'_U + Q'_U : U = c + H_{\mathrm{NE}}, c \in \mathbb{Z}^2, A_h \in \sigma^{\mathrm{op}}_-(A) \right\}.$$

The reverse inclusion "⊃" is checked in an analogous way as in the proof of Proposition 3.89, using the fact that A is rich.

The operators in $\sigma_{NE}^{op}(\oplus A_\tau)$ and $\sigma_{NW}^{op}(\oplus A_\tau)$ are exactly the shifts $V_c B V_{-c}$ of the operators B in (iv) and (v) with A_h replaced by the operator $A_{c^{(1)}+h}$, respectively, which is in the same local operator spectrum as A_h is, by Proposition 3.94.   □

This proposition, in combination with the equivalence of properties (A) and (C) above, leaves us with a stability criterion for the finite section method of a certain class of operators A on E.

4.1.2 The Main Theorem on the Finite Section Method

To give us a more concise notation, abbreviate

$$P := P_{\mathbb{D}_+} \quad \text{and} \quad Q := I - P,$$

which are the projection to the positive half axis and its complementary projection, respectively.

Theorem 4.2. *If $A \in \mathrm{BDO}_{S,\$}^p$, and either*

(i) $E = \ell^p(\mathbb{Z}, \mathbf{X})$ *with $p \in [1, \infty]$, a Banach space $\mathbf{X}$, and $\mathbb{D} = \mathbb{Z}$, or*

(ii) $E = L^p$ *with $p \in (1, \infty)$ and $\mathbb{D} = \mathbb{R}$, or*

(iii) $E = L^p$ *with $p \in \{1, \infty\}$, $A \in \mathcal{R}_p$, and $\mathbb{D} = \mathbb{R}$,*

then the following statements are equivalent.

① *The finite section method (4.1) is applicable to A.*

② *The finite section sequence (4.1) is stable.*

③ *A itself is invertible, and the set consisting of all operators $\oplus_{\tau \in \mathbb{D}}(Q V_\tau A_h V_{-\tau} Q + P)$ with $A_h \in \sigma_+^{op}(A)$ and all $\oplus_{\tau \in \mathbb{D}}(P V_\tau A_h V_{-\tau} P + Q)$ with $A_h \in \sigma_-^{op}(A)$ is uniformly invertible.*

④ *A is invertible, and all sets $\{Q V_\tau A_h V_{-\tau} Q + P\}_{\tau \in \mathbb{D}}$ with $A_h \in \sigma_+^{op}(A)$ and $\{P V_\tau A_h V_{-\tau} P + Q\}_{\tau \in \mathbb{D}}$ with $A_h \in \sigma_-^{op}(A)$ are essentially invertible, and the essential suprema of all these inverses are uniformly bounded.*

In case (i), properties ①, ②, ③ and ④ are moreover equivalent to

⑤ *A is invertible, and the set consisting of all operators*

$$Q A_h Q + P \quad \text{with} \quad A_h \in \sigma_+^{op}(A) \tag{4.3}$$

as well as all

$$P A_h P + Q \quad \text{with} \quad A_h \in \sigma_-^{op}(A) \tag{4.4}$$

is uniformly invertible.

Proof. We start with the proof of the equivalence of ② and ③. By Theorem 2.28 or 2.47, depending on whether case (i), (ii) or (iii) applies, we get that $(A_{\lceil\tau\rfloor})$ is stable if and only if $\oplus A_{\lceil\tau\rfloor}$ is invertible at infinity, which, by Proposition 4.1 and Theorem 1, is equivalent to the uniform invertibility of $\sigma^{\mathrm{op}}(\oplus A_{\lceil\tau\rfloor})$. By Proposition 4.1, this is equivalent to the uniform invertibility of the sets

$$\{\oplus(V_{-c}AV_c) \;:\; c \in \mathbb{D}\}, \tag{4.5}$$

$$\{\oplus A_h \;:\; c \in \mathbb{D}\}, \tag{4.6}$$

$$\{P'_{\mathsf{NE}}(\oplus A_h)P'_{\mathsf{NE}} + Q'_{\mathsf{NE}} \;:\; A_h \in \sigma^{\mathrm{op}}_-(A)\}, \tag{4.7}$$

$$\text{and } \{P'_{\mathsf{NW}}(\oplus A_h)P'_{\mathsf{NW}} + Q'_{\mathsf{NW}} \;:\; A_h \in \sigma^{\mathrm{op}}_+(A)\}, \tag{4.8}$$

where all four $\oplus$ signs in (4.5)–(4.8) are over $\tau \in \mathbb{D}$, meaning that every one of these four stacked operators is the same in every layer.

Firstly, the uniform invertibility of (4.5) implies the invertibility of A, which, secondly, implies the uniform invertibility of the sets (4.5) and (4.6), by Proposition 2.20 and Theorem 1.

From

$$P'_{\mathsf{NE}} = \bigoplus_{\tau\in\mathbb{D}} (V_{-\tau}PV_\tau) \qquad \text{and} \qquad Q'_{\mathsf{NE}} = \bigoplus_{\tau\in\mathbb{D}} (V_{-\tau}QV_\tau)$$

we get that the operators in (4.7) are of the form

$$\bigoplus_{\tau\in\mathbb{D}} \left(V_{-\tau}(PV_\tau A_h V_{-\tau}P + Q)V_\tau \right).$$

By Proposition 2.20, and since V_τ and $V_{-\tau}$ are invertible isometries, this operator is invertible if and only if

$$\bigoplus_{\tau\in\mathbb{D}} (PV_\tau A_h V_{-\tau}P + Q)$$

is invertible, where the norms of their inverses coincide. Consequently, the uniform invertibility of (4.7) is equivalent to the uniform invertibility of the first set in ③. Analogously, one checks that (4.8) corresponds to the second set in ③, and we are finished with the equivalence of ② and ③.

The equivalence of ③ and ④ is immediate from Proposition 2.20.

Since ② is equivalent to ③, it implies the invertibility of A. But then, by Proposition 1.78, ① and ② are equivalent.

Finally, let $\mathbb{D} = \mathbb{Z}$. Then the word "essential" becomes "uniform" in ④, by the discussion after (2.10), and the union of all sets discussed in ④ is exactly the set of operators in (4.3) and (4.4), by Proposition 3.94, which proves the equivalence of ④ and ⑤ if $\mathbb{D} = \mathbb{Z}$. □

Remark 4.3. a) If $p \in \{1, \infty\}$ or $A \in \mathcal{W}$, we know from Section 3.9 that the uniform invertibility of $\sigma^{\mathrm{op}}(\oplus A_{\lceil\tau\rfloor})$ is guaranteed by its elementwise invertibility. That is why, in these cases, the word "uniformly" can be replaced by "elementwise" in ③, and the uniformity condition in ④ is redundant as well.

b) In ③ and ④ it is obviously not necessary to distinguish between limit operators A_h and $A_{h'}$ if h' is of the form $h + d$ for some $d \in \mathbb{Z}$.

c) The operators in (4.3) and (4.4) are the operators $(A_h)_-$ and $(A_h)_+$ in the notation of (3.14). We will make use of this observation in the next subsection.

d) Note that ⑤ is in general not equivalent to ①–④ if $\mathbb{D} = \mathbb{R}$, as the following example shows. However, note that ⑤ were equivalent to ①–④ if we had defined limit operators of A on L^p with respect to sequences $h = (h_m) \subset \mathbb{R}^n$. (See Subsection 3.4.13 why we did not.)

Example 4.4. For every $k \in \mathbb{Z}$, let J_k denote the flip on the interval $(k, k+1)$; that is

$$(J_k u)(x) = \begin{cases} u(2k + 1 - x) & x \in (k, k + 1), \\ 0 & \text{otherwise,} \end{cases}$$

and put

$$A := \sum_{k \in \mathbb{Z}} J_k,$$

the sum in $(Au)(x)$ actually being finite for every $x \in \mathbb{R}$.

Then a finite section $A_{\lceil \tau \rceil}$ is invertible if and only if τ is an integer, and hence, the finite section sequence $(A_{\lceil \tau \rceil})$ is far away from being stable. But A is (integer-) translation invariant and thus coincides with all of its limit operators A_h. So it is easy to see that property ⑤ is fulfilled!   □

4.1.3   Two Baby Versions of Theorem 4.2

Before we apply our theorem in a more sophisticated setting in the next section, we will study two cases in which Theorem 4.2 takes an especially simple form in this short subsection.

Translation Invariant Operators

A very simple case is that of a $\mathbb{D}$-translation invariant operator; that is an operator $A \in \mathrm{BDO}^p_{\mathcal{S},\$}$ with $V_{-\tau} A V_\tau = A$ for all $\tau \in \mathbb{D}$.

Corollary 4.5. *Let the conditions of Theorem 4.2 be fulfilled, and suppose $A \in \mathrm{BDO}^p_{\mathcal{S},\$}$ is $\mathbb{D}$-translation invariant. Then the finite section method (4.1) is applicable to A if and only if A itself and the two operators A_+ and A_- are invertible, where, as agreed in (3.14),*

$$A_+ = PAP + Q \qquad and \qquad A_- = QAQ + P.$$

Proof. This follows immediately from the equivalence of ① and ④ in Theorem 4.2 since A coincides with all of its limit operators A_h and their shifts $V_\tau A_h V_{-\tau}$ with $\tau \in \mathbb{D}$ if A is $\mathbb{D}$-translation invariant.   □

The Discrete Case with Slowly Oscillating Coefficients

Suppose $E = \ell^p = \ell^p(\mathbb{Z}, \mathbb{C})$ with $1 < p < \infty$; that is $n = 1$, $\mathbf{X} = \mathbb{C}$ and $\mathbb{D} = \mathbb{Z}$. We will study the finite section method for band-dominated operators A on E with slowly oscillating coefficients.

For simplicity, we start with a band operator A on E with slowly oscillating coefficients. In that case, we have the following simplification of Theorem 4.2.

Corollary 4.6. *If $A \in$ BO has slowly oscillating coefficients and $\mathbb{D} = \mathbb{Z}$, then the finite section method (4.1) is applicable to A if and only if A and all operators (4.3) and (4.4) are invertible.*

Proof. Indeed, recalling Remark 4.3 a), the uniform boundedness condition in ④ is redundant since $\oplus A_\tau \in$ BO $\subset W$ by Proposition 2.22 a). Moreover, the sets in ④ are all singletons since the limit operators of A are translation invariant by Remark 3.53. With these modifications, ④ is equal to ⑤ with "uniformly" replaced by "elementwise", and our proof is finished. □

Remark 4.7. a) Note that Proposition 3.113 is not applicable here to remove the uniform boundedness condition in ④ of Theorem 4.2 since although $A \in$ BDO^p has slowly oscillating coefficients, the same is in general not true for $\oplus A_\tau$.

b) In fact, the statement of Corollary 4.6 is even true for arbitrary $A \in W$ without the premise of slowly oscillating coefficients. This can be shown by replacing the stacked operator $\oplus A_\tau$ with the block diagonal operator

$$\text{diag}(P_1 A P_1 , P_2 A P_2 , \dots)$$

acting on the same space E as A does. For example, see [69] for the adjustment of the proofs to this setting. The reason why we didn't present this idea here is that it is limited to $\mathbb{D} = \mathbb{Z}$ only, while our aim was to present a unified approach for continuous and discrete approximation methods.

So, besides the invertibility of A, which is an indispensable ingredient to the applicability of any approximation method to A of course, all we need for the applicability of the finite section method is the invertibility of all operators

$$B_- \text{ with } B \in \sigma_+^{\text{op}}(A) \qquad \text{and} \qquad C_+ \text{ with } C \in \sigma_-^{\text{op}}(A), \qquad (4.9)$$

where B_- and C_+ are defined by (3.14). From Corollary 3.19 we know that the invertibility of A implies that B_- and C_+ are Fredholm, and

$$\text{ind } B_- = -\text{ind}^+ A \qquad \text{and} \qquad \text{ind } C_+ = -\text{ind}^- A = \text{ind}^+ A \qquad (4.10)$$

for all $B \in \sigma_+^{\text{op}}(A)$ and $C \in \sigma_-^{\text{op}}(A)$. Since all coefficients of A are slowly oscillating, the limit operators B and C are Laurent operators, by Remark 3.53. Consequently, B_- and C_+ are Toeplitz operators on the respective half axes, and for those we have Coburn's theorem (Theorem 2.38 in [9] or Theorem 1.10 in [10]) saying that $\text{ind } B_- = 0$ and $\text{ind } C_+ = 0$ already imply the invertibility of B_- and C_+.

Summarizing this with (4.10) and the fact that (4.1) is applicable if and only if A and all operators (4.9) are invertible, we get:

Proposition 4.8. *If $A \in BO$ has slowly oscillating coefficients and $\mathbb{D} = \mathbb{Z}$, then the finite section method (4.1) is applicable to A if and only if A is invertible and* $\mathrm{ind}^+ A = 0$.

In [74], ROCH generalized this result to band-dominated operators with slowly oscillating coefficients, so that Proposition 4.8 is still true if we replace BO by BDO^p. Moreover, ROCH's result extends from $1 < p < \infty$ to $A \in \mathrm{BDO}^1$ and $A \in \mathrm{BDO}^\infty$, provided $A \in \mathcal{W}$. Note that this is a generalization of the classical result on the stability of the finite section method for band-dominated Laurent operators, which is the case of constant coefficients.

4.2   The FSM for a Class of Integral Operators

For the remainder of this book, we will dive into the continuous case, where we study the finite section method in an algebra containing a huge amount of operators emerging from practical problems.

4.2.1   An Algebra of Convolution and Multiplication Operators

Let $E = L^p = L^p(\mathbb{R}^n)$ with some $p \in [1, \infty]$ and $n \in \mathbb{N}$. With every function $\kappa \in L^1$, we associate the operator that maps a function $u \in E$ to the convolution $\kappa \star u$; that is the function

$$(\kappa \star u)(x) = \int_{\mathbb{R}^n} \kappa(x - y)\, u(y)\, dy, \qquad x \in \mathbb{R}^n.$$

From Young's inequality [72] we know that $\kappa \star u \in E$ and $\|\kappa \star u\| \leq \|\kappa\|_1 \|u\|$, so that $u \mapsto \kappa \star u$ is a bounded linear operator on E.

For reasons which will become apparent in a minute, it is more convenient to associate the operator $u \mapsto \kappa \star u$ with the Fourier transform

$$(F\kappa)(x) = \int_{\mathbb{R}^n} \kappa(y)\, e^{i\langle x, y\rangle}\, dy, \qquad x \in \mathbb{R}^n$$

of κ rather than with κ, where $\langle \cdot, \cdot \rangle$ is the standard scalar product in $\mathbb{R}^n$. The *convolution operator* $u \mapsto \kappa \star u$ is henceforth denoted by C_a, where $a = F\kappa$ is the Fourier transform of κ, which is also referred to as the *symbol* of C_a. We denote the set of functions $\{F\kappa : \kappa \in L^1\}$ by FL^1. From what we said earlier, we know that $C_a \in L(E)$ and $\|C_a\| \leq \|\kappa\|_1$ for every $a = F\kappa \in FL^1$.

From the associativity $\kappa_1 \star (\kappa_2 \star u) = (\kappa_1 \star \kappa_2) \star u$ and the basic convolution theorem $F(\kappa_1 \star \kappa_2) = (F\kappa_1)(F\kappa_2)$ we get that

$$C_{a_1} C_{a_2} = C_{F(\kappa_1 \star \kappa_2)} = C_{(F\kappa_1)(F\kappa_2)} = C_{a_1 a_2},$$

where $a_1 = F\kappa_1$ and $a_2 = F\kappa_2$.

Moreover, it can be shown that the spectrum $\operatorname{sp} C_a$ coincides with the range $a(\dot{\mathbb{R}}^n)$ of the symbol $a \in FL^1$, where one easily verifies that $FL^1 \subset C(\dot{\mathbb{R}}^n)$ with $a(\infty) = 0$ for every $a \in FL^1$.

Therefore, C_a is never invertible if $a \in FL^1$, while an operator $\lambda I - C_a$ is invertible if and only if $\lambda \notin a(\dot{\mathbb{R}}^n)$. We will widen our class of convolution operators by saying that $\lambda I + C_a$ is the convolution operator with symbol $b = \lambda + a = \lambda + F\kappa$ and denote this operator by C_b; that is

$$(C_b u)(x) = \lambda u(x) + \int_{\mathbb{R}^n} \kappa(x - y)\, u(y)\, dy, \qquad x \in \mathbb{R}^n$$

for all $\lambda \in \mathbb{C}$ and $\kappa \in L^1$. Moreover, we write $\mathbb{C} + FL^1$ as an abbreviation for the set $\{\lambda + F\kappa : \lambda \in \mathbb{C}, \kappa \in L^1\}$. Note that this enlargement of FL^1 is equivalent to the enlargement of L^1 by the delta distribution and its multiples $\lambda\delta$ with $\lambda \in \mathbb{C}$.

It follows immediately from what we said earlier that $C_{a_1} C_{a_2} = C_{a_1 a_2}$ and $\operatorname{sp} C_a = a(\dot{\mathbb{R}}^n)$ also hold for all symbols a_1, a_2 and a in $\mathbb{C} + FL^1$, where $a(\infty) = \lambda$ if $a = \lambda + F\kappa \in \mathbb{C} + FL^1 \subset C(\dot{\mathbb{R}}^n)$.

The study of convolution operators can be pretty interesting in its own right but the class of operators to be studied enjoys a lot more variety if also multiplication operators enter the scene.

Depending on the application that we have in mind, choose some Banach subalgebra Y of L^∞, and let $\mathcal{A}_Y^o$ and $\mathcal{A}_Y$ denote the smallest subalgebra and Banach subalgebra, respectively, of $L(E)$ containing all convolution operators C_a with $a \in FL^1$ and all multiplication operators M_b with $b \in Y$; that is

$$\mathcal{A}_Y^o := \operatorname{alg}_{L(E)}\big\{ C_a, M_b : a \in FL^1,\ b \in Y \big\},$$
$$\mathcal{A}_Y := \operatorname{clos}_{L(E)} \mathcal{A}_Y^o.$$

If $Y = L^\infty$, then the subscript Y will be omitted.

We start our studies of the Banach algebra $\mathcal{A}_Y$ by a simple observation, where we say that an operator $A \in L(E)$ is *locally compact* if $P_U A$ and $A P_U$ are compact operators for all bounded and measurable sets $U \subset \mathbb{R}^n$, which is clearly equivalent to the discretization (1.3) being locally compact in the sense of Definition 2.14.

Lemma 4.9. *Convolution operators C_a with symbol $a \in FL^1$ are locally compact.*

Proof. Let $a = F\kappa$ with $\kappa \in L^1$, and let $U \subset \mathbb{R}^n$ be bounded and measurable. Since κ can be approximated in the norm of L^1 as well as desired by some compactly supported continuous function κ' and since

$$\|C_{F\kappa} - C_{F\kappa'}\| = \|C_{F(\kappa - \kappa')}\| \leq \|\kappa - \kappa'\|_1, \tag{4.11}$$

it suffices to prove our claim with κ' in place of κ.

Therefore, let $V \subset \mathbb{R}^n$ denote a compact set which contains U as well as the support of κ'. Then $P_V C_{F\kappa'}$ is an integral operator with a continuous kernel function supported in $(2V) \times V$. But since such operators are compact, also

$$P_U C_{F\kappa'} = P_U(P_V C_{F\kappa'})$$

is compact. The proof of the second claim is completely symmetric.   □

This fact has remarkable consequences as we will see. By its definition, the algebra $\mathcal{A}_Y^o$ consists of all finite sums of the form

$$A = \sum_{i=1}^{k} A_i,$$

where A_i are finite products of convolution and multiplication operators. Some of these A_i might be pure multiplication operators, say $A_1, \ldots, A_j$, and all other terms, $A_{j+1}, \ldots, A_k$ involve at least one convolution operator. The sum $A_1 + \cdots + A_j$ is again a multiplication operator whose symbol shall be denoted by $b^A \in Y$. So

$$A = M_{b^A} + \sum_{i=j+1}^{k} A_i. \tag{4.12}$$

This additive decomposition in a pure multiplication operator and a remaining term involving convolution operators is uniquely determined. We will show that, in a sense, this is even true for arbitrary $A \in \mathcal{A}_Y$. Therefore, recall that

$$\mathcal{J}_Y := \mathrm{closid}_{\mathcal{A}_Y}\{C_a : a \in FL^1\}$$

is the smallest closed ideal in $\mathcal{A}_Y$ containing all convolution operators with symbol in FL^1. Again, we omit the subscript Y if $Y = L^\infty$. It is an elementary exercise to show that

$$\mathcal{J}_Y = \mathrm{clos}_{L(E)}\left\{ \sum_i A_i C_{a_i} B_i \ : \ A_i, B_i \in \mathcal{A}_Y^o, \ a_i \in FL^1 \right\}, \tag{4.13}$$

the finite summation being over all terms of the given form.

Lemma 4.10. *All operators in $\mathcal{J}_Y$ are locally compact.*

Proof. Take an arbitrary bounded and measurable set $U \subset \mathbb{R}^n$. From Lemma 4.9 and the fact that P_U commutes with every multiplication operator we get that $P_U A C_a$ and $C_a B P_U$ are compact for all $A, B \in \mathcal{A}_Y^o$ and all $a \in FL^1$. The rest follows from formula (4.13).   □

Proposition 4.11. *The decomposition $\mathcal{A}_Y = \{M_b : b \in Y\} \dotplus \mathcal{J}_Y$ holds.*

Proof. Clearly, on the right we have two closed subalgebras of $\mathcal{A}_Y$. Suppose A is contained in both. Then, for every bounded and measurable set $U \subset \mathbb{R}^n$, the operator $P_U A$ is compact by Lemma 4.10. This implies $b = 0$, and hence $A = 0$.

Let us now show that the sum on the right is all of $\mathcal{A}_Y$. Every operator in the (dense) subset $\mathcal{A}_Y^o$ can be written as such a sum, by (4.12). It remains to show that the (linear) mapping $A \mapsto M_{b^A}$ is bounded on $\mathcal{A}_Y^o$, and hence can be continuously extended to $\mathcal{A}_Y$.

To see this, take some $A \in \mathcal{A}_Y^o$, an $\varepsilon > 0$ and a bounded and measurable set $U \subset \mathbb{R}^n$ with positive measure such that $\|P_U b^A\|_\infty > \|b^A\|_\infty - \varepsilon$. By (4.12) and Lemma 4.10, we have $P_U A - P_U M_{b^A} \in K(E)$. Now

$$
\begin{aligned}
\|b^A\|_\infty - \varepsilon \;\leq\; & \|P_U M_{b^A}\| = \|P_U M_{b^A} + K(E)\|_{L(E)/K(E)} \\
= \; & \|P_U A + K(E)\|_{L(E)/K(E)} \;\leq\; \|P_U A\| \;\leq\; \|A\|
\end{aligned}
$$

for every $\varepsilon > 0$, and hence, $\|M_{b^A}\| = \|b^A\|_\infty \leq \|A\|$. $\qquad\square$

The continuous extension of the mapping $A \mapsto M_{b^A}$ to all of $\mathcal{A}_Y$ shall be denoted by μ. The bounded linear mapping $\mu : \mathcal{A}_Y \to \mathcal{A}_Y$ is obviously a projection from $\mathcal{A}_Y$ onto $\{M_b : b \in Y\}$. The complementary projection $A \mapsto A - M_{b^A}$ from $\mathcal{A}_Y$ onto $\mathcal{J}_Y$ shall be denoted by ϱ. The decomposition

$$
A \;=\; \mu(A) \;+\; \varrho(A) \tag{4.14}
$$

of $A \in \mathcal{A}_Y$ into a multiplication part and a locally compact part is the generalization of the simple idea (4.12) to the Banach algebra $\mathcal{A}_Y$.

Proposition 4.11 allows us to identify the factor algebra $\mathcal{A}_Y/\mathcal{J}_Y$ with the algebra $\{M_b : b \in Y\}$, and hence with $Y \subset L^\infty$, where the multiplication operator $\mu(A)$ is a representative of the coset $A + \mathcal{J}_Y$ for every $A \in \mathcal{A}_Y$.

Proposition 4.12. a) *If $A \in \mathcal{A}$ is invertible at infinity and the multiplication operator $\mu(A)$ is invertible, then A is Fredholm.*

b) *If $A \in \mathcal{A}$ is Fredholm and $1 < p < \infty$, then A is invertible at infinity.*

Proof. a) is an immediate consequence of the decomposition (4.14), Lemma 4.10 and Proposition 2.15, and b) follows from Figure 4 on page 57. $\qquad\square$

The condition $\kappa \in L^1$ is sufficient and necessary for C_{Fk} being in the Wiener algebra. Consequently, we have the following.

Proposition 4.13. *Every operator $A \in \mathcal{A}^o$ is in the Wiener algebra $\mathcal{W}$.*

Proof. Of course, every multiplication operator is in $\mathcal{W}$, and it is verified in almost complete analogy to Example 1.45 that $C_a \in \mathcal{W}$ for all $a \in FL^1$. The claim then follows from the closedness of $\mathcal{W}$ under finite sums and products. $\qquad\square$

4.2.2 The Finite Section Method in $\mathcal{A}_\$$

We will now study the applicability of the finite section method for operators in the smallest closed subalgebra

$$
\mathcal{A}_\$ \;:=\; \mathcal{A}_{L_\$^\infty}
$$

of $L(E)$ containing all convolutions C_a with $a \in FL^1$ and all multiplications M_b by a rich function $b \in L_{\$}^{\infty}$. To be able to apply our Theorem 4.2, we put $n = 1$, and hence $E = L^p := L^p(\mathbb{R})$ with some $p \in [1, \infty]$.

Proposition 4.14. *Every operator $A \in \mathcal{A}_{\$}$ is subject to the conditions on Theorem 4.2; that is $A \in \mathrm{BDO}_{\mathcal{S}, \$}^p$ and, if $p \in \{1, \infty\}$, then $A \in \mathcal{R}_p$.*

Proof. By Proposition 3.11 a), for the proof of $\mathcal{A}_{\$} \subset \mathrm{BDO}_{\mathcal{S}, \$}^p$ it is sufficient to show that the generators C_a and M_b of our algebra $\mathcal{A}_{\$}$ are in $\mathrm{BDO}_{\mathcal{S}, \$}^p$.

Let $a = F\kappa$ with $\kappa \in L^1$. Since $\|P_m\kappa - \kappa\|_1 \to 0$ as $m \to \infty$, we get that $\|C_{a_m} - C_a\| \to 0$ as $m \to \infty$, where $a_m := F(P_m\kappa)$; that is, we replace the convolution kernel κ by its truncation to the interval $[-m, m]$. But from $C_{a_m} \in \mathrm{BO}$ for all $m \in \mathbb{N}$, we get that $C_a \in \mathrm{BDO}^p$. Alternatively, also Proposition 4.13 immediately shows that $C_a \in \mathrm{BDO}^p$. Obviously, also $M_b \in \mathrm{BDO}^p$ for all $b \in L_{\$}^{\infty}$.

If $p = \infty$, taking into account that the pre-adjoint operators of C_a and M_b are $C_{\tilde{a}}$ and M_b, respectively, where $\tilde{a}(t) = a(-t)$, we clearly have $C_a, M_b \in \mathcal{S}$.

Convolution operators C_a are translation invariant, and hence rich, and our multiplication operators M_b are rich by our premise $b \in L_{\$}^{\infty}$.

It remains to show that $\mathcal{A}_{\$} \subset \mathcal{R}_p$ if $p \in \{1, \infty\}$. We start with $p = 1$ and take an arbitrary $A \in \mathcal{A}_{\$}$. We have to show that $A_{\lceil \tau \rfloor} \in \mathbf{r}(L^1)$ for every $\tau > 0$. Therefore write $A = M_b + J$ with $M_b = \mu(A)$ and $J = \varrho(A) \in \mathcal{J}_{L_{\$}^{\infty}}$ as in (4.14). Then we have

$$A_{\lceil \tau \rfloor} = P_\tau A P_\tau + Q_\tau = P_\tau M_b P_\tau + P_\tau J P_\tau + Q_\tau = M_{b_\tau} + K,$$

where $b_\tau = P_\tau b + Q_\tau 1$ and $K = P_\tau J P_\tau \in K(E)$ by Lemma 4.10. What we have to show is that this operator is subject to (2.16).

So suppose $A_{\lceil \tau \rfloor}$ is bounded below. Then, by Lemma 2.32, it is a Φ_+-operator, and this set is closed under compact perturbations, by Lemma 2.41. So M_{b_τ} is in $\Phi_+(L^1)$ as well. Lemma 2.42 shows that then M_{b_τ} is invertible on L^1. Consequently, $A_{\lceil \tau \rfloor} = M_{b_\tau} + K$, as a compact perturbation of an invertible operator, is a Fredholm operator of index zero. On the other hand, it has kernel dimension zero, by Lemma 2.32, and so it is invertible.

For the proof of $\mathcal{A}_{\$} \subset \mathcal{R}_p$ for $p = \infty$, we just note that $A^\triangleleft \in \mathcal{A}_{\$}$ in L^1 if $A \in \mathcal{A}_{\$}$ in L^∞ since the pre-adjoint operators of C_a and M_b are $C_{\tilde{a}}$ and M_b, respectively, where $\tilde{a}(t) = a(-t)$. $\square$

The Slowly Oscillating Case

We can now apply Theorem 4.2 to operators in our algebra $\mathcal{A}_{\$}$ to get some sufficient and necessary criteria for the applicability of their finite section method. We demonstrate this for the case $\mathcal{A}_{\mathrm{SO}}^o$, which is contained in $\mathcal{A}_{\$}$ by Proposition 3.49, where we get a nice and very explicit result.

By Proposition 3.47, the limit operator of an operator of multiplication by a slowly oscillating function is just a multiple of the identity operator. Applying

Proposition 3.4, we get that every limit operator of $A \in \mathcal{A}_{SO}$ is in

$$\text{closalg}_{L(E)}\{C_a, I \,:\, a \in FL^1\} = \{C_a \,:\, a \in \mathbb{C} + FL^1\}.$$

We collect all these symbols $a \in \mathbb{C} + FL^1$ of limit operators A_h towards $+\infty$ in a set $\mathcal{R}_+$ and those towards $-\infty$ in $\mathcal{R}_-$. So we have $\mathcal{R}_\pm \subset \mathbb{C} + FL^1$ with

$$\sigma_\pm^{\text{op}}(A) = \{C_a \,:\, a \in \mathcal{R}_\pm\},$$

respectively. Finally, put $\mathcal{R} := \mathcal{R}_+ \cup \mathcal{R}_-$.

Proposition 4.15. *For $A \in \mathcal{A}_{SO}^0$, the following statements are equivalent.*

(i) *The finite section method (4.1) is applicable to A.*

(ii) *The finite section sequence (4.1) is stable.*

(iii) *The operator A itself is invertible, and for all functions $a \in \mathcal{R}$, the range $a(\dot{\mathbb{R}})$ is a closed curve in the complex plane which does not contain the origin and has winding number zero.*

Proof. By their definition, all functions $a \in \mathcal{R}$ are in $\mathbb{C} + FL^1$, and hence, they are continuous on the one-point compactification $\dot{\mathbb{R}}$, i.e. each range $a(\dot{\mathbb{R}})$ is some closed curve in the complex plane.

Proposition 4.14 and Theorem 4.2 yield that (i) and (ii) are equivalent to each other and to the uniform invertibility of the set consisting of

(1)   the operator A itself,

(2)   all $\underset{\tau \in \mathbb{R}}{\oplus} (QV_\tau A_h V_{-\tau} Q + P)$ with A_h running through $\sigma_+^{\text{op}}(A)$, and

(3)   all $\underset{\tau \in \mathbb{R}}{\oplus} (PV_\tau A_h V_{-\tau} P + Q)$ with A_h running through $\sigma_-^{\text{op}}(A)$.

By Propositions 4.13 and 3.112, the word "uniform" can be replaced by "elementwise" in the previous sentence.

We start by examining the operators in (3). Since $A_h \in \sigma_-^{\text{op}}(A)$, we have $A_h = C_a$ with some $a \in \mathcal{R}_-$. Hence, A_h is translation invariant and

$$PV_\tau A_h V_{-\tau} P + Q = PC_a P + Q =: W_a$$

for every $\tau \in \mathbb{R}$. The operators W_a are *Wiener-Hopf operators* (recall Subsection 3.7.2). Such an operator W_a is invertible if and only if the image of its symbol a does not contain the origin and has winding number zero with respect to the origin (see, for instance, [29]). Analogously, we get that (2) is equivalent to the invertibility of all $QC_aQ + P$ with $a \in \mathcal{R}_+$. But $QC_aQ + P$ is invertible if and only if $W_a = PC_aP + Q$ is invertible.

So all operators in (2) and (3) are invertible if and only if the symbols of all functions $a \in \mathcal{R}$ stay away from the origin and have winding number zero. $\qquad\square$

Remark 4.16. Remember that every function $b \in L^\infty$ which converges at $+\infty$ and $-\infty$, is contained in SO. If all multiplication operators involved in A have this property, then the sets $\mathcal{R}_+$ and $\mathcal{R}_-$ are just singletons, their union $\mathcal{R}$ has at most two elements, and the number of curves to be examined in property (*iii*) of Proposition 4.15 is (at most) two. So the criterion for applicability of the FSM is very effective in this case, indeed. This special case was already studied, using slightly different methods, in [46].

Note that, in this setting, Proposition 4.15 also yields the classical results of GOHBERG and FELDMAN [29] about the applicability of projection methods to paired equations; that is

$$A = PC_a + QC_b \quad \text{or} \quad A = C_a P + C_b Q$$

with $a, b \in \mathbb{C} + FL^1$, the discrete analogues of which we already encountered in (3.50) and (3.51).

4.2.3   A Special Finite Section Method for BC

In this section we will study an approximation method for operators on the space of bounded and continuous functions $\mathrm{BC} \subset L^\infty$. The operators of interest to us will be of the form

$$A = I + K,$$

where K shall be bounded and linear on L^∞ with the condition $Ku \in \mathrm{BC}$ for all $u \in L^\infty$. Typically, K will be some integral operator. One of the simplest examples is a convolution operator $K = C_a$ with some $a \in FL^1$. In this simple case, the validity of the above condition can be seen as follows.

Lemma 4.17. *If $a \in FL^1$, then $C_a u$ is a continuous function for every $u \in L^\infty$.*

Proof. Let $a = F\kappa$ with some $\kappa \in L^1$. Since κ can be approximated in the norm of L^1 as close as desired by a continuous function with a compact support, we may, by (4.11), suppose that κ already is such a function. But in this case,

$$\left| (C_a u)(x_1) - (C_a u)(x_2) \right| = \left| \int_{\mathbb{R}^n} \Big(\kappa(x_1 - y) - \kappa(x_2 - y) \Big) u(y) \, dy \right|$$

$$\leq \|u\|_\infty \int_{\mathbb{R}^n} |\kappa(t + \Delta x) - \kappa(t)| \, dt$$

clearly tends to zero as $\Delta x = x_1 - x_2 \to 0$. $\qquad\square$

As a slightly more sophisticated example one could look at an operator of the following form or at the norm limit of a sequence of such operators.

Example 4.18. Put

$$K := \sum_{i=1}^{j} M_{b_i} C_{a_i} M_{c_i}, \qquad (4.15)$$

where $b_i \in BC$, $a_i \in FL^1$, $c_i \in L^\infty$ and $j \in \mathbb{N}$. For the condition that K maps L^∞ into BC, it is sufficient to impose continuity of the functions b_i in (4.15), whereas the functions c_i need not be continuous since their action is smoothed by the convolution thereafter. □

The following auxiliary result is essentially Lemma 3.4 from [4]. Since we will apply this result in different situations in what follows, we pass to different notations for a moment. Let L be bounded and linear on L^∞ with im $L \subset BC$, and put $B = I + L$. Moreover, abbreviate the restriction $B|_{BC}$ by B_0.

Lemma 4.19. a) *$Bu \in BC$ if and only if $u \in BC$.*

b) *B is invertible on L^∞ if and only if B_0 is invertible on BC. In this case*

$$\|B_0^{-1}\|_{L(BC)} \leq \|B^{-1}\|_{L(L^\infty)} \leq 1 + \|B_0^{-1}\|_{L(BC)} \|L\|_{L(L^\infty)}. \qquad (4.16)$$

Proof. a) This is immediate from $Bu = u + Lu$ and $Lu \in BC$ for all $u \in L^\infty$.

b) If B is invertible on L^∞, then the invertibility of B_0 on BC and the first inequality in (4.16) follows from a).

Now let B_0 be invertible on BC. To see that B is injective on L^∞, suppose $Bu = 0$ for $u \in L^\infty$. From $0 \in BC$ and a) we get that $u \in BC$ and hence $u = 0$ since B is injective on BC. Surjectivity of B on L^∞: Since B_0 is surjective on BC, for every $v \in L^\infty$ there is a $u \in BC$ such that $B_0 u = Lv \in BC$. Consequently,

$$B(v - u) = Bv - B_0 u = v + Lv - Lv = v \qquad (4.17)$$

holds, showing the surjectivity of B on L^∞. So B is invertible on L^∞, and, by (4.17), $B^{-1}v = v - u = v - B_0^{-1}Lv$ for all $v \in L^\infty$, and hence $B^{-1} = I - B_0^{-1}L$. This proves the second inequality in (4.16). □

Remark 4.20. a) The previous lemma clearly holds for arbitrary Banach spaces with one of them contained in the other in place of BC and L^∞.

b) If, moreover, L has a pre-adjoint operator on L^1, then an approximation argument as in the proof of Lemma 3.5 of [4] even shows that, in fact, (4.16) can be improved to the equality $\|B_0^{-1}\|_{L(BC)} = \|B^{-1}\|_{L(L^\infty)}$.

We are looking for solutions u of the equation $Au = b$, which typically is some integral equation

$$u(x) + \int_{\mathbb{R}^n} k(x,y)\, u(y)\, dy = b(x), \qquad x \in \mathbb{R}^n, \qquad (4.18)$$

for any right hand side $b \in BC$. By Lemma 4.19 a) with $L = K$ we are looking for u in BC only.

In this setting, a popular approximation method is just to reduce the range of integration from $\mathbb{R}^n$ to $[-\tau, \tau]^n$ which is clearly different from the FSM. We call this procedure the *finite section method for* BC, short: BC-FSM. We are now looking for solutions $u_\tau \in$ BC of

$$u_\tau(x) + \int_{[-\tau,\tau]^n} k(x,y)\, u_\tau(y)\, dy \;=\; b(x), \qquad x \in \mathbb{R}^n \tag{4.19}$$

with $\tau > 0$, and hope that the sequence (u_τ) of solutions of (4.19) $\mathcal{P}$-converges to the solution u of (4.18) as $\tau \to \infty$. Note that, unlike in our previous methods of the form (1.34), this time the right hand side b remains unchanged, and the support of u_τ is not limited to some cube!

The modified finite section method (4.19) can be written as $A_\tau u_\tau = b$ with

$$A_\tau \;=\; I + K P_\tau. \tag{4.20}$$

As a consequence of Lemma 4.19 a) with $L = K P_\tau$ one also has

Corollary 4.21. *For every $\tau > 0$, it holds that $A_\tau u_\tau \in$ BC if and only if $u_\tau \in$ BC.*

In contrast to the usual finite section method (4.1) on L^∞, the method (4.20) is taylor-made for operators of the form $A = I + K$ acting on BC. We can regard A_τ (or better, its restriction to BC) as an element of $L(\mathrm{BC})$, and we have $u_\tau \in$ BC for all $\tau > 0$ if $b \in$ BC, whereas another projector P_τ from the left in (4.20) would obviously spoil this setting.

In accordance with the machinery presented up to this point, our strategy to study equation (4.18) and the stability of its approximation by (4.19) is to embed these into L^∞, where we can relate the BC-FSM (4.19) to our usual FSM (4.1) on L^∞. Indeed, the applicabilities of these different methods turn out to be equivalent.

Proposition 4.22. *For the operator $A = I + K$ with $\operatorname{im} K \subset$ BC,*

$$A_\tau \;=\; I + K P_\tau \qquad \text{and} \qquad A_{[\tau]} \;=\; P_\tau A P_\tau + Q_\tau, \qquad \tau > 0,$$

the following statements are equivalent.

(i) *The BC-FSM (A_τ) alias (4.20) is applicable in* BC.

(ii) *The BC-FSM (A_τ) alias (4.20) is applicable in L^∞.*

(iii) *The finite section method $(A_{[\tau]})$ is applicable in L^∞.*

(iv) *(A_τ) is stable on* BC.

(v) *(A_τ) is stable on L^∞.*

(vi) *$(A_{[\tau]})$ is stable on L^∞.*

Proof. The implication (i)$\Rightarrow$(iv) is standard. The equivalence of (iv) and (v) follows from Lemma 4.19 b) with $L = K P_\tau$. The equivalence of (v) and (vi) comes from the following observation.

$$
\begin{aligned}
A_\tau = I + KP_\tau &= P_\tau + P_\tau KP_\tau + Q_\tau + Q_\tau KP_\tau \\
&= P_\tau AP_\tau + Q_\tau(I + Q_\tau KP_\tau) \\
&= (P_\tau AP_\tau + Q_\tau)(I + Q_\tau KP_\tau) \\
&= A_{\lceil \tau \rceil}(I + Q_\tau KP_\tau),
\end{aligned}
$$

where the second factor $(I + Q_\tau KP_\tau)$ is always invertible with its inverse equal to $I - Q_\tau KP_\tau$, and hence $\|(I + Q_\tau KP_\tau)^{-1}\| \le 1 + \|K\|$ for all $\tau > 0$.

(v)$\Rightarrow$(ii). Since (v) implies (vi), it also implies the invertibility of A on L^∞ by Theorem 4.2. But this, together with (v), implies (ii) by Theorem 1.75.

Finally, the implication (ii)$\Rightarrow$(i) is trivial if we keep in mind Lemma 4.19 a) and Corollary 4.21, and the equivalence of (iii) and (vi) follows from Theorem 4.2. $\qquad\square$

For the study of property (iii) in Proposition 4.22 we have Theorem 4.2 involving limit operators of A, provided that, in addition, A is a rich operator.

In Example 4.18 we therefore require the functions b_i and c_i to be rich. By Proposition 3.39, this is equivalent to $b_i \in \mathrm{BUC} = \mathrm{BC} \cap L_{\$}^\infty$ and $c_i \in L_{\$}^\infty$. In the particularly simple case when $n = 1$ and all functions b_i and c_i have limits $b_i^\pm, c_i^\pm \in \mathbb{C}$ at plus and minus infinity, respectively, this gives us that the statements of Proposition 4.22 are equivalent to the operator A and two associated Wiener-Hopf operators being invertible. These two Wiener-Hopf operators are the compression of $I + C_{a^+}$ to the negative half axis and that of $I + C_{a^-}$ to the positive half axis, where

$$
a^\pm = \sum_{i=1}^{j} b_i^\pm c_i^\pm a_i \in FL^1,
$$

respectively. They are invertible if and only if the closed curves $a^+(\mathbb{R})$ and $a^-(\mathbb{R})$ do not contain the point $-1 \in \mathbb{C}$ and have winding number zero with respect to it. In Subsection 4.3.4 we will study similar problems when $A = I + K$ is a boundary integral operator in so-called rough surface problems.

Remark 4.23. – Other approximation methods. Although we have restricted ourselves to the FSM (4.1) and the BC-FSM (4.20), there are some other approximation methods in BC which are "close enough" to the FSM to handle the question of their applicability in terms of our results on the finite section method.

For example, instead of the finite section projectors P_τ one can consider projections P^k, $k \in \mathbb{N}$, onto spaces of polynomial splines. These splines shall be defined on a uniform mesh on the cube $[-\tau(k), \tau(k)]^n$ with increasing density as k increases and with $\tau(k) \to \infty$ as $k \to \infty$.

As in Chapter 5 of [58] it can be seen that, after a sensible choice of these spline subspaces, the resulting Galerkin method $(P^k AP^k)_{k=1}^\infty$ is close enough to the finite section method $(P_{\tau(k)} AP_{\tau(k)})_{k=1}^\infty$ for its applicability being equivalent to that of the latter method.

4.3   Boundary Integral Equations on Unbounded Surfaces

A large number of important problems in physical sciences and engineering can be modelled by a strongly elliptic boundary value problem on a domain D; that is a strongly elliptic PDE in the domain, coupled with an additional condition at the boundary ∂D. A common technique to study such problems is to reformulate the boundary value problem as a boundary integral equation [53, 79].

The theory of the boundary integral equation method when the boundary ∂D is compact and smooth enough (at least Lipschitz) is very well developed, for example, see [53, 79, 34, 25]. The theory is much less well-developed in the case when ∂D is unbounded; clearly, additional issues arise in numerical solution, in particular the issue of truncation of ∂D (involving the finite section method).

In this section we consider the case where

$$D = \{(x, z) \in \mathbb{R}^{n+1} : x \in \mathbb{R}^n , z > f(x)\} \tag{4.21}$$

is the epigraph of a given bounded and continuous function $f : \mathbb{R}^n \to \mathbb{R}$. Let

$$f_+ = \sup_{x \in \mathbb{R}^n} f(x) \qquad \text{and} \qquad f_- = \inf_{x \in \mathbb{R}^n} f(x)$$

denote the highest and the lowest elevation of the infinite surface ∂D. It is convenient to assume, without loss of generality, that $f_- > 0$, so that D is entirely contained in the half space $H = \{(x, z) \in \mathbb{R}^{n+1} : x \in \mathbb{R}^n , z > 0\}$.

There is a lot of ongoing research for problems on such domains. A number of specific problems of this type have been rigorously formulated as boundary value problems and reformulated as second kind integral equations. For example, see [22, 21, 20, 16] for acoustic problems, [2, 3] for problems involving elastic waves, and [57] for a boundary value problem for Laplace's equation arising in the study of unsteady water waves. A small selection of literature on the numerical analysis of these problems is [15, 16, 19, 20, 42].

4.3.1   The Structure of the Integral Operators Involved

The integral operators arising in all these reformulations are of the form

$$(Ku)(x) = \int_{\mathbb{R}^n} k(x, y) \, u(y) \, dy, \qquad x \in \mathbb{R}^n \tag{4.22}$$

for $u \in BC$ with

$$k(x, y) = \sum_{i=1}^{j} b_i(x) \, k_i \big(x - y, f(x), f(y)\big) \, c_i(y), \tag{4.23}$$

where

$$b_i \in BC, \quad k_i \in C\big((\mathbb{R}^n \setminus \{0\}) \times [f_- , f_+]^2\big) \quad \text{and} \quad c_i \in L^\infty \tag{4.24}$$

for $i = 1, \ldots, j$, and

$$|k(x, y)| \leq \kappa(x - y), \qquad x, y \in \mathbb{R}^n, \tag{4.25}$$

for some $\kappa \in L^1$. We denote the set of all operators K satisfying (4.22)–(4.25) for a particular function $f \in \mathrm{BC}$ by $\mathcal{K}_f$.

Remark 4.24. a) Inequality (4.25) ensures that K acts boundedly on L^∞ and, moreover, that it is band-dominated, and even contained in the Wiener algebra. This is reflected by the methods used in the engineering literature [14, 86].

b) Note that, as is the case in the following example, inequality (4.25) need not hold for $k_i(x - y, f(x), f(y))$ in place of $k(x, y)$.

Example 4.25. – Unsteady water waves. In [57] PRESTON, CHAMBERLAIN and CHANDLER-WILDE consider the two-dimensional Dirichlet boundary value problem: Given $\varphi_0 \in \mathrm{BC}(\partial D)$, find $\varphi \in C^2(D) \cap \mathrm{BC}(\overline{D})$ such that

$$\begin{aligned} \triangle\varphi &= 0 & \text{in} \quad D, \\ \varphi &= \varphi_0 & \text{on} \quad \partial D, \end{aligned}$$

which arises in the theory of classical free surface water wave problems. In this case, $n = 1$ and the surface function f shall in addition be differentiable with an α-Hölder continuous first derivative for a fixed $\alpha \in (0, 1]$; precisely, for some constant $C > 0$, $|f'(x) - f'(y)| \leq C|x - y|^\alpha$ holds for all $x, y \in \mathbb{R}^n$.

Now let

$$G(\mathbf{x}, \mathbf{y}) = \Phi(\mathbf{x}, \mathbf{y}) - \Phi(\mathbf{x}^r, \mathbf{y}), \qquad \mathbf{x}, \mathbf{y} \in \mathbb{R}^2$$

denote the Green's function in the half plane H where

$$\Phi(\mathbf{x}, \mathbf{y}) = -\frac{1}{2\pi} \ln |\mathbf{x} - \mathbf{y}|_2, \qquad \mathbf{x}, \mathbf{y} \in \mathbb{R}^2,$$

with $|\,.\,|_2$ denoting the Euclidean norm in $\mathbb{R}^2$, is the fundamental solution for Laplace's equation in two dimensions, and $\mathbf{x}^r = (x_1, -x_2)$ is the reflection of $\mathbf{x} = (x_1, x_2)$ with respect to ∂H. For the solution of the above boundary value problem we make the following double layer potential ansatz:

$$\varphi(\mathbf{x}) = \int_{\partial D} \frac{\partial G(\mathbf{x}, \mathbf{y})}{\partial \mathbf{n}(\mathbf{y})} \tilde{u}(\mathbf{y}) \, ds(\mathbf{y}), \qquad \mathbf{x} \in D,$$

where $\mathbf{n}(\mathbf{y}) = (f'(y), -1)$ denotes the normal vector of ∂D at $\mathbf{y} = (y, f(y))$, and we look for the corresponding density function $\tilde{u} \in \mathrm{BC}(\partial D)$. In [57] it is shown that φ satisfies the above Dirichlet boundary value problem if and only if

$$(I - K)\tilde{u} = -2\varphi_0, \tag{4.26}$$

where

$$(K\tilde{u})(\mathbf{x}) = 2 \int_{\partial D} \frac{\partial G(\mathbf{x}, \mathbf{y})}{\partial \mathbf{n}(\mathbf{y})} \tilde{u}(\mathbf{y}) \, ds(\mathbf{y}), \qquad \mathbf{x} \in \partial D.$$

In accordance with the parametrization $\mathbf{x} = (x, f(x))$ of ∂D, we define

$$u(x) := \tilde{u}(\mathbf{x}) \quad \text{and} \quad b(x) := -2\varphi_0(\mathbf{x}), \quad x \in \mathbb{R}$$

and rewrite equation (4.26) as the equation

$$u(x) - \int_{-\infty}^{+\infty} k(x, y) \, u(y) \, dy = b(x), \quad x \in \mathbb{R} \tag{4.27}$$

on the real axis for the unknown function $u \in \mathrm{BC}(\mathbb{R})$, where

$$
\begin{aligned}
k(x, y) &= 2\frac{\partial G(\mathbf{x}, \mathbf{y})}{\partial \mathbf{n}(\mathbf{y})} \sqrt{1 + f'(y)^2} = -\frac{1}{\pi}\left(\frac{(\mathbf{x} - \mathbf{y}) \cdot \mathbf{n}(\mathbf{y})}{|\mathbf{x} - \mathbf{y}|_2^2} - \frac{(\mathbf{x}^r - \mathbf{y}) \cdot \mathbf{n}(\mathbf{y})}{|\mathbf{x}^r - \mathbf{y}|_2^2}\right) \\
&= -\frac{1}{\pi}\left(\frac{(x - y)f'(y) - f(x) + f(y)}{(x - y)^2 + (f(x) - f(y))^2} - \frac{(x - y)f'(y) + f(x) + f(y)}{(x - y)^2 + (-f(x) - f(y))^2}\right) \\
&= -\frac{1}{\pi}\left(\frac{x - y}{(x - y)^2 + (f(x) - f(y))^2} - \frac{x - y}{(x - y)^2 + (f(x) + f(y))^2}\right)f'(y) \\
&\quad + \frac{1}{\pi}\left(\frac{f(x) - f(y)}{(x - y)^2 + (f(x) - f(y))^2} + \frac{f(x) + f(y)}{(x - y)^2 + (f(x) + f(y))^2}\right)
\end{aligned}
$$

is clearly of the form (4.23) with $j = 2$ and property (4.24) satisfied. From Lemma 2.1 and inequality (5) in [57] we moreover get that inequality (4.25) holds with

$$\kappa(x) = \begin{cases} c|x|^{\alpha-1} & \text{if } 0 < |x| \leq 1, \\ c|x|^{-2} & \text{if } |x| > 1, \end{cases}$$

where $\alpha \in (0, 1]$ is the Hölder exponent of f', and c is some positive constant. $\square$

Example 4.26. – Wave scattering on an unbounded rough surface. In [21] CHANDLER-WILDE, ROSS and ZHANG consider the corresponding problem to the preceding Example 4.25 for the Helmoltz equation in two dimensions:

Given $\varphi_0 \in \mathrm{BC}(\partial D)$, they seek $\varphi \in C^2(D) \cap \mathrm{BC}(\overline{D})$ such that

$$
\begin{aligned}
\Delta\varphi + k^2\varphi &= 0 \quad \text{in} \quad D, \\
\varphi &= \varphi_0 \quad \text{on} \quad \partial D,
\end{aligned}
$$

and such that φ satisfies an appropriate radiation condition and constraints on growth at infinity. Again, $n = 1$ and the surface function f shall be differentiable with an α-Hölder continuous first derivative for some $\alpha \in (0, 1]$.

This problem models the scattering of acoustic waves by a sound-soft rough surface; the same problem arises in time-harmonic electromagnetic scattering by a perfectly conducting rough surface. The constant k in the Helmholtz equation is the wave number.

The authors reformulate this problem as a boundary integral equation which has exactly the form (4.26), where $G(\mathbf{x}, \mathbf{y})$ is now defined to be the Green's function for Helmholtz's instead of Laplace's equation in the half plane H which satisfies the impedance condition $\partial G/\partial x_2 + ikG = 0$ on ∂H. As in Example 4.25, this boundary integral equation can be written in the form (4.27) with $k(x,y)$ of the form (4.23) with $j = 2$ and property (4.24) satisfied, and also here inequality (4.25) holds with

$$\kappa(x) = \begin{cases} c\,|x|^{\alpha-1} & \text{if } 0 < |x| \leq 1, \\ c\,|x|^{-3/2} & \text{if } |x| > 1, \end{cases}$$

where $\alpha \in (0, 1]$ is the Hölder exponent of f', and c is some positive constant. $\square$

Example 4.27. – **Wave propagation over a flat inhomogeneous surface.** The propagation of mono-frequency acoustic or electromagnetic waves over flat inhomogeneous terrain has been modelled in two dimensions by the Helmholtz equation

$$\triangle\varphi + k^2\varphi = 0$$

in the upper half plane $D = H$ (so $f \equiv 0$ in (4.21)) with a Robin (or impedance) condition

$$\frac{\partial\varphi}{\partial x_2} + ik\beta\varphi = \varphi_0$$

on the boundary line ∂D. Here k, the wavenumber, is constant, $\beta \in L^\infty(\partial D)$ is the surface admittance describing the local properties of the ground surface ∂D, and the inhomogeneous term φ_0 is in $L^\infty(\partial D)$ as well.

Similarly to Example 4.26, in fact using the same Green's function $G(\mathbf{x}, \mathbf{y})$ of the Helmholtz equation, CHANDLER-WILDE, RAHMAN and ROSS in [20] reformulate this problem as a boundary integral equation on the real line,

$$u(x) \; - \; \int_{-\infty}^{+\infty} \tilde{\kappa}(x - y)z(y)u(y)\,dy = \psi(x), \qquad x \in \mathbb{R}, \qquad (4.28)$$

where $\psi \in BC$ is given and $u \in BC$ is to be determined. The function $\tilde{\kappa}$ is in L^1, and $z \in L^\infty$ is closely connected with the surface admittance β by $z = i(1 - \beta)$.

Note that the kernel function of the integral operator in (4.28) is of the form (4.23) with $j = 1$. The validity of (4.24) and (4.25) is trivial in this case. $\square$

For technical reasons we find it convenient to embed the class $\mathcal{K}_f$ of integral operators (4.22) with a kernel $k(\cdot, \cdot)$ subject to (4.23), (4.24) and (4.25) into a somewhat larger Banach algebra of integral operators[1]. Therefore, given $f \in BC$, put $f_- := \inf f$, $f_+ := \sup f$, and let R_f denote the set of all operators of the form

$$(Bu)(x) = \int_{\mathbb{R}^n} k\big(x - y, f(x), f(y)\big)\,u(y)\,dy, \qquad x \in \mathbb{R}^n \qquad (4.29)$$

---

[1]It will turn out that this Banach algebra actually is not that much larger than our class $\mathcal{K}_f$.

with $k \in C(\mathbb{R}^n \times [f_-, f_+]^2)$ compactly supported, and put

$$
\begin{aligned}
\hat{\mathcal{B}} &:= \text{closspan}\{ M_b B M_c : b \in \text{BC}, B \in R_f, c \in L^\infty \}, \\
\mathcal{B} &:= \text{closalg}\{ M_b B M_c : b \in \text{BC}, B \in R_f, c \in L^\infty \}, \\
\hat{\mathcal{C}} &:= \text{closspan}\{ M_b C_a M_c : b \in \text{BC}, a \in FL^1, c \in L^\infty \}, \\
\mathcal{C} &:= \text{closalg}\{ M_b C_a M_c : b \in \text{BC}, a \in FL^1, c \in L^\infty \}.
\end{aligned}
$$

Remark 4.28. a) Here, closspan M denotes the closure in $L(\text{BC})$ of the set of all finite sums of elements of $M \subset L(\text{BC})$, and, as agreed before, closalg M denotes the closure in $L(\text{BC})$ of the set of all finite sum-products of elements of M. So closspan M is the smallest closed subspace and closalg M the smallest (not necessarily unital) Banach subalgebra of $L(\text{BC})$ containing M. In both cases we say they are *generated by* M.

b) The following proposition shows that $\hat{\mathcal{B}}$ and $\mathcal{B}$ do not depend on the function $f \in \text{BC}$ which is why we omit f in their notations.

c) It is easily seen that all operators in $\hat{\mathcal{C}}$ map arbitrary elements from L^∞ into BC. Consequently, every $K \in \hat{\mathcal{C}}$ is subject to the condition on K in Subsection 4.2.3.

d) The linear space $\hat{\mathcal{C}}$ is the closure of the set of operators covered by Example 4.18. The following proposition shows that this set already contains all of $\mathcal{K}_f$. More precisely, it coincides with the closure of $\mathcal{K}_f$ in the norm of $L(\text{BC})$ and with the other spaces and algebras introduced above.

Proposition 4.29. *The identity*

$$
\text{clos}\,\mathcal{K}_f = \hat{\mathcal{B}} = \hat{\mathcal{C}} = \mathcal{B} = \mathcal{C} \subset \mathcal{A}
$$

holds.

Proof. Clearly, $\hat{\mathcal{C}} \subset \hat{\mathcal{B}}$ since C_a with $a \in FL^1$ can be approximated in the operator norm by convolutions $B = C_{a'}$ with a continuous and compactly supported kernel. But these operators B are clearly in R_f.

For the reverse inclusion, $\hat{\mathcal{B}} \subset \hat{\mathcal{C}}$, it is sufficient to show that the generators of $\hat{\mathcal{B}}$ are contained in $\hat{\mathcal{C}}$. We will prove this by showing that $B \in \hat{\mathcal{C}}$ for all $B \in R_f$. So let $k \in C(\mathbb{R}^n \times [f_-, f_+]^2)$ be compactly supported, and put B as in (4.29). To see that $B \in \hat{\mathcal{C}}$, take $L \in \mathbb{N}$, choose $f_- = s_1 < s_2 < \cdots < s_{L-1} < s_L = f_+$ equidistant in $[f_-, f_+]$, and let φ_ξ denote the standard Courant hat function for this mesh that is centered at s_ξ. Then, since k is uniformly continuous, its piecewise linear interpolations (with respect to the variables s and t),

$$
k^{(L)}(r, s, t) := \sum_{\xi,\eta=1}^{L} k(r, s_\xi, s_\eta)\,\varphi_\xi(s)\,\varphi_\eta(t), \qquad r \in \mathbb{R}^n,\ s,t \in [f_-, f_+],
$$

uniformly approximate k as $L \to \infty$, whence the corresponding integral operators with k replaced by $k^{(L)}$ in (4.29),

$$(B^{(L)}u)(x) = \int_{\mathbb{R}^n} \sum_{\xi,\eta=1}^{L} k(x-y, s_\xi, s_\eta) \, \varphi_\xi(f(x)) \, \varphi_\eta(f(y)) \, u(y) \, dy, \qquad (4.30)$$

converge to B in the operator norm as $L \to \infty$. But it is obvious from (4.30) that $B^{(L)} \in \hat{\mathcal{C}}$, which proves that also $B \in \hat{\mathcal{C}}$.

To see that $\mathcal{B} = \mathcal{C}$, it is sufficient to show that the generators of each of the algebras are contained in the other algebra. But this follows from $\hat{\mathcal{B}} = \hat{\mathcal{C}}$, which is already proven.

That $\mathcal{C}$ is contained in the Banach algebra $\mathcal{A}$ generated by L^1-convolutions and L^∞-multiplications, is obvious.

For the inclusion $\mathcal{C} \subset \hat{\mathcal{C}}$ it is sufficient to show that $C_a M_b C_c \in \hat{\mathcal{C}}$ for all $a, c \in FL^1$ and $b \in L^\infty$. So take an arbitrary $b \in L^\infty$ and let $a = F\kappa$ and $c = F\lambda$ with $\kappa, \lambda \in L^1$. Without loss of generality, by (4.11), we may suppose that κ and λ are continuous and compactly supported, say $\kappa(x) = \lambda(x) = 0$ if $|x| > \ell$. It is now easily checked that

$$(C_a M_b C_c u)(x) = \int_{\mathbb{R}^n} k(x, y) u(y) \, dy, \qquad x \in \mathbb{R}^n$$

with

$$k(x, y) = \int_{\mathbb{R}^n} \kappa(x-z) \, b(z) \, \lambda(z-y) \, dz = \int_{|t| \le \ell} \kappa(t) \, b(x-t) \, \lambda(x-t-y) \, dt.$$

By taking a sufficiently fine partition $\{T_1, \ldots, T_N\}$ of $[-\ell, \ell]^n$ and fixing $t_m \in T_m$ for $m = 1, \ldots, N$, we can approximate the above arbitrarily close by

$$k(x, y) \;=\; \sum_{m=1}^{N} \int_{T_m} \kappa(t) \, b(x-t) \, \lambda(x-t-y) \, dt$$

$$\approx \sum_{m=1}^{N} \kappa(t_m) \, \lambda(x-t_m-y) \int_{T_m} b(x-t) \, dt$$

$$= \sum_{m=1}^{N} \kappa_m \, \lambda_m(x-y) \, b_m(x), \qquad x, y \in \mathbb{R}^n \qquad (4.31)$$

where $\kappa_m = \kappa(t_m)$, $\lambda_m(x) = \lambda(x-t_m)$ and $b_m(x) = \int_{T_m} b(x-t) \, dt$, the latter depending continuously on x. But this clearly shows that $C_a M_b C_c \in \hat{\mathcal{C}}$.

The inclusion $\hat{\mathcal{B}} \subset \operatorname{clos} \mathcal{K}_f$ is also obvious since (4.24) and (4.25) hold if $b_i \in BC$, $c_i \in L^\infty$ and k_i is compactly supported and continuous on all of $\mathbb{R}^n \times [f_-, f_+]^2$.

So it remains to show that $\operatorname{clos} \mathcal{K}_f \subset \hat{\mathcal{B}}$. This clearly follows if we show that $\mathcal{K}_f \subset \hat{\mathcal{B}}$. So let $K \in \mathcal{K}_f$ be arbitrary, that means K is an integral operator of the form (4.22) with a kernel $k(.,.)$ subject to (4.23), (4.24) and (4.25). For every $\ell \in \mathbb{N}$, let $p_\ell : [0, \infty) \to [0, 1]$ denote a continuous function with support in $[1/(2\ell), 2\ell]$ which is identically equal to 1 on $[1/\ell, \ell]$. Then, for $i = 1, \ldots, j$,

$$k_i^{(\ell)}(r, s, t) := p_\ell(|r|) \, k_i(r, s, t), \qquad r \in \mathbb{R}^n, \ s, t \in [f_-, f_+]$$

is compactly supported and continuous on $\mathbb{R}^n \times [f_-, f_+]^2$, whence $B_i^{(\ell)} \in R$ with

$$(B_i^{(\ell)} u)(x) := \int_{\mathbb{R}^n} k_i^{(\ell)}(x - y, f(x), f(y)) \, u(y) \, dy, \qquad x \in \mathbb{R}^n$$

for all $u \in BC$. Now put

$$
\begin{aligned}
k^{(\ell)}(x, y) &:= \sum_{i=1}^j b_i(x) \, k_i^{(\ell)}(x - y, f(x), f(y)) \, c_i(y) \\
&= p_\ell(|x - y|) \, k(x, y),
\end{aligned}
$$

and let $K^{(\ell)}$ denote the operator (4.22) with k replaced by $k^{(\ell)}$; that is

$$K^{(\ell)} = \sum_{i=1}^j M_{b_i} B_i^{(\ell)} M_{c_i}, \tag{4.32}$$

which is clearly in $\hat{\mathcal{B}}$. It remains to show that $K^{(\ell)} \rightrightarrows K$ as $\ell \to \infty$. Therefore, note that

$$
\begin{aligned}
\|K - K^{(\ell)}\| &\leq \sup_{x \in \mathbb{R}^n} \int_{\mathbb{R}^n} \left| k(x, y) - k^{(l)}(x, y) \right| \, dy \\
&= \sup_{x \in \mathbb{R}^n} \int_{\mathbb{R}^n} \left| \left(1 - p_\ell(|x - y|) \right) k(x, y) \right| \, dy \\
&\leq \sup_{x \in \mathbb{R}^n} \int_{|x-y|<1/\ell} |k(x, y)| \, dy + \sup_{x \in \mathbb{R}^n} \int_{|x-y|>\ell} |k(x, y)| \, dy \\
&\leq \sup_{x \in \mathbb{R}^n} \int_{|x-y|<1/\ell} |\kappa(x - y)| \, dy + \sup_{x \in \mathbb{R}^n} \int_{|x-y|>\ell} |\kappa(x - y)| \, dy \\
&= \int_{|z|<1/\ell} |\kappa(z)| \, dz + \int_{|z|>\ell} |\kappa(z)| \, dz
\end{aligned}
$$

with $\kappa \in L^1$ from (4.25). But clearly, this goes to zero as $\ell \to \infty$. $\qquad \square$

4.3.2   Limit Operators of these Integral Operators

In order to apply our previous results on the finite section method for $A = I + K$, we need to know about the limit operators of A, which, clearly, reduces to finding the limit operators of $K \in \mathcal{K}_f$. But before we start looking for these limit operators, we single out a subclass $\mathcal{K}_f^{\$}$ of $\mathcal{K}_f$ all elements of which are rich operators. So, this time given $f \in \mathrm{BUC}$, let

$$
\begin{aligned}
\mathcal{K}_f^{\$} &:= \{ K \in \mathcal{K}_f : b_i \in \mathrm{BUC},\ c_i \in L_{\mathrm{SC\$}}^\infty \text{ for } i = 1, \dots, j \}, \\
\hat{\mathcal{B}}_{\$} &:= \mathrm{closspan}\{ M_b B M_c : b \in \mathrm{BUC},\ B \in R_f,\ c \in L_{\mathrm{SC\$}}^\infty \}, \\
\mathcal{B}_{\$} &:= \mathrm{closalg}\{ M_b B M_c : b \in \mathrm{BUC},\ B \in R_f,\ c \in L_{\mathrm{SC\$}}^\infty \}, \\
\hat{\mathcal{C}}_{\$} &:= \mathrm{closspan}\{ M_b C_a M_c : b \in \mathrm{BUC},\ a \in FL^1,\ c \in L_{\mathrm{SC\$}}^\infty \}, \\
\mathcal{C}_{\$} &:= \mathrm{closalg}\{ M_b C_a M_c : b \in \mathrm{BUC},\ a \in FL^1,\ c \in L_{\mathrm{SC\$}}^\infty \}
\end{aligned}
$$

denote the rich counterparts of $\mathcal{K}_f, \hat{\mathcal{B}}, \mathcal{B}, \hat{\mathcal{C}}$ and $\mathcal{C}$, and put

$$
\mathcal{A}_{\$}' := \mathrm{closalg}\{ M_b,\ C_a M_c : b \in L_{\$}^\infty,\ a \in FL^1,\ c \in L_{\mathrm{SC\$}}^\infty \} \supset \mathcal{A}_{\$}.
$$

Recall that, by Proposition 3.39, $\mathrm{BC} \cap L_{\$}^\infty = \mathrm{BUC}$, and that $C_a M_c$ is rich for all $a \in FL^1$ and $c \in L_{\mathrm{SC\$}}^\infty$ by Proposition 3.79, whence every operator in $\mathcal{A}_{\$}'$ is rich. Then the following "rich version" of Proposition 4.29 holds.

Proposition 4.30. *If $f \in \mathrm{BUC}$, then it holds that*

$$
\mathrm{clos}\,\mathcal{K}_f^{\$} = \hat{\mathcal{B}}_{\$} = \hat{\mathcal{C}}_{\$} = \mathcal{B}_{\$} = \mathcal{C}_{\$} \subset \mathcal{A}_{\$}'.
$$

In particular, every $K \in \mathcal{K}_f^{\$}$ is rich.

Proof. All we have to check is that the arguments we made in the proof of Proposition 4.29 preserve membership of b and c in BUC and $L_{\mathrm{SC\$}}^\infty$, respectively. In only two of these arguments there are multiplications by b and c involved at all.

The first one is the proof of the inclusion $\hat{\mathcal{B}} \subset \hat{\mathcal{C}}$. In this argument, we show that every $B \in R_f$ is contained in $\hat{\mathcal{C}}$. But in fact, this construction even yields $B \in \hat{\mathcal{C}}_{\$}$, which can be seen as follows. $B \in R_f$ is approximated in the operator norm by the operators $B^{(L)}$ from (4.30). Since the Courant hats φ_ξ and φ_η are in BUC and also $f \in \mathrm{BUC}$, we get $\varphi_\xi \circ f \in \mathrm{BUC}$ and $\varphi_\eta \circ f \in \mathrm{BUC} \subset L_{\mathrm{SC\$}}^\infty$. So $B^{(L)} \in \hat{\mathcal{C}}_{\$}$, and hence $B \in \hat{\mathcal{C}}_{\$}$.

The second argument involving multiplication operators is the proof of the inclusion $\mathcal{C} \subset \hat{\mathcal{C}}$. But also at this point it is easily seen that the functions $b_m(x) = \int_{T_m} b(x - t)\, dt$ that are invoked in (4.31) are in fact in BUC, whence $\mathcal{C}_{\$} \subset \hat{\mathcal{C}}_{\$}$. $\square$

Now we are ready to say something about the limit operators of $K \in \mathcal{K}_f^{\$}$. Not surprisingly, the key to these operators is the behaviour of the surface function f and of the multipliers b_i and c_i at infinity. We will show that every limit operator K_h of K is of the same form (4.22) but with f, b_i and c_i replaced by $f^{(h)}$, $b_i^{(h)}$ and $\tilde{c}_i^{(h)}$, respectively, in (4.23). We will even formulate and prove the analogous result for operators in $\mathcal{B}_{\$}$. The key step to this result is the following lemma.

Lemma 4.31. *Let* $B \in R_f$; *that is,* B *is of the form* (4.29) *with a compactly supported kernel function* $k \in C(\mathbb{R}^n \times [f_-, f_+]^2)$, *and let* $c \in L^\infty_{\text{SCS}}$. *If a sequence* $h = (h_m) \subset \mathbb{Z}^n$ *tends to infinity and the functions* $f^{(h)}$ *and* $\tilde{c}^{(h)}$ *exist, then the limit operator* $(BM_c)_h$ *exists and is the integral operator*

$$\Big((BM_c)_h u \Big)(x) = \int_{\mathbb{R}^n} k\big(x - y, f^{(h)}(x), f^{(h)}(y)\big)\, \tilde{c}^{(h)}(y)\, u(y)\, dy, \qquad x \in \mathbb{R}^n. \tag{4.33}$$

Proof. Choose $\ell > 0$ large enough that $k(r, s, t) = 0$ for all $r \in \mathbb{R}^n$ with $|r| \geq \ell$ and all $s, t \in [f_-, f_+]$. Now take a sequence $h = (h_m) \subset \mathbb{Z}^n$ such that the functions $f^{(h)}$ and $\tilde{c}^{(h)}$ exist. By definition of $f^{(h)}$ and $\tilde{c}^{(h)}$, this is equivalent to

$$\big\| f|_{h_m + U} - f^{(h)}|_U \big\|_\infty \to 0 \qquad \text{and} \qquad \big\| c|_{h_m + U} - \tilde{c}^{(h)}|_U \big\|_1 \to 0 \tag{4.34}$$

as $m \to \infty$ for every compactum $U \subset \mathbb{R}^n$. Moreover, it is easily seen that

$$(V_{-h_m} BM_c V_{h_m} u)(x) = \int_{\mathbb{R}^n} k\big(x - y, f(x + h_m), f(y + h_m)\big)\, c(y + h_m)\, u(y)\, dy$$

for all $x \in \mathbb{R}^n$ and $u \in BC$. Abbreviating $A_m := V_{-h_m} BM_c V_{h_m} - (BM_c)_h$, we get that $(A_m u)(x) = \int_{\mathbb{R}^n} d_m(x, y)\, u(y)\, dy$, where

$$|d_m(x, y)|$$
$$= \Big| k\big(x - y, f(x + h_m), f(y + h_m)\big)\, c(y + h_m)$$
$$\qquad - k\big(x - y, f^{(h)}(x), f^{(h)}(y)\big)\, \tilde{c}^{(h)}(y) \Big|$$
$$\leq \Big| k\big(x - y, f(x + h_m), f(y + h_m)\big) - k\big(x - y, f^{(h)}(x), f^{(h)}(y)\big) \Big| \cdot \|c\|_\infty$$
$$\qquad + \|k\|_\infty \cdot \Big| c(y + h_m) - \tilde{c}^{(h)}(y) \Big| \tag{4.35}$$

for all $x, y \in \mathbb{R}^n$ and $m \in \mathbb{N}$. Moreover, $d_m(x, y) = 0$ if $|x - y| \geq \ell$.

Now take an arbitrary $\tau > 0$ and put $U := [-\tau - \ell, \tau + \ell]^n$ and $V := [-\tau, \tau]^n$. Then, by (4.35),

$$\|P_\tau A_m\| = \text{ess sup}_{x \in V} \int_{\mathbb{R}^n} |d_m(x, y)|\, dy = \text{ess sup}_{x \in V} \int_U |d_m(x, y)|\, dy \to 0$$

as $m \to \infty$ since (4.34) holds and k is uniformly continuous. Analogously,

$$\|A_m P_\tau\| = \text{ess sup}_{x \in \mathbb{R}^n} \int_V |d_m(x, y)|\, dy = \text{ess sup}_{x \in U} \int_V |d_m(x, y)|\, dy \to 0$$

as $m \to \infty$. This proves that $(BM_c)_h$ from (4.33) is indeed the limit operator of BM_c with respect to the sequence $h = (h_m)$. $\square$

Proposition 4.32.   a) *Let $K = M_b B M_c$ with $b \in \mathrm{BUC}$, $B \in R_f$ and $c \in L^\infty_{\mathrm{SC\$}}$. If $h = (h_m) \subset \mathbb{Z}^n$ tends to infinity and all functions $b^{(h)}$, $f^{(h)}$ and $\tilde{c}^{(h)}$ exist, then the limit operator K_h exists and is the integral operator*

$$(K_h u)(x) = \int_{\mathbb{R}^n} b^{(h)}(x) \, k\big(x{-}y, f^{(h)}(x), f^{(h)}(y)\big) \, \tilde{c}^{(h)}(y) \, u(y) \, dy, \qquad x \in \mathbb{R}^n. \tag{4.36}$$

b) *Every limit operator of $K = M_b B M_c$ with $b \in \mathrm{BUC}$, $B \in R_f$ and $c \in L^\infty_{\mathrm{SC\$}}$ is of this form (4.36).*

c) *The mapping $K \mapsto K_h$ acting on $\{M_b B M_c : b \in \mathrm{BUC}, \; B \in R_f, \; c \in L^\infty_{\mathrm{SC\$}}\}$ as given in (4.36) extends to a continuous Banach algebra homomorphism on all of $\mathcal{B}_\$$ by passing to an appropriate subsequence of h if required. In particular, all limit operators K_h of $K \in \mathcal{K}_f \subset \mathcal{B}$ are of the form (4.22) with k replaced by*

$$\hat{k}^{(h)}(x, y) = \sum_{i=1}^{j} b_i^{(h)}(x) \, k_i\big(x - y, f^{(h)}(x), f^{(h)}(y)\big) \, \tilde{c}_i^{(h)}(y). \tag{4.37}$$

Proof. a) From Proposition 3.4 we get that K_h exists and is equal to $(M_b)_h (B M_c)_h$ which is exactly (4.36) by Lemma 4.31.

b) Suppose $g \subset \mathbb{Z}^n$ is a sequence tending to infinity that leads to a limit operator K_g of K. Since $b, f \in L^\infty$ and $c \in L^\infty_{\mathrm{SC\$}}$, there is a subsequence h of g such that the functions $b^{(h)}$, $f^{(h)}$ and $\tilde{c}^{(h)}$ exist. But then we are in the situation of a), and the limit operator K_h of K exists and is equal to (4.36). Since h is a subsequence of g, we have $K_g = K_h$.

c) The extension to $\mathcal{B}_\$$ follows from Proposition 3.4. The formula for the limit operators of $K \in \mathcal{K}_f$ follows from the approximation of K by (4.32) for which we explicitly know the limit operators. $\qquad\square$

Example 4.33. Suppose $K \in \mathcal{K}_f$ where the surface function f and the functions b_i and c_i are all slowly oscillating. Let $h \subset \mathbb{Z}^n$ be a sequence tending to infinity such that $b_i^{(h)}$, $f^{(h)}$ and $\tilde{c}_i^{(h)}$ exist – otherwise pass to a subsequence of h with this property which is always possible.

From Corollary 3.48 we know that all of $b_i^{(h)}$, $f^{(h)}$ and $\tilde{c}_i^{(h)} = c_i^{(h)}$ are constant. Then, by Proposition 4.32 c), the limit operator K_h is the integral operator with kernel function

$$\hat{k}^{(h)}(x, y) = \sum_{i=1}^{j} b_i^{(h)} \, \tilde{c}_i^{(h)} \, k_i\big(x - y, f_h^{(h)}, f_h^{(h)}\big), \qquad x, y \in \mathbb{R}^n \tag{4.38}$$

which is just a pure operator of convolution by $\hat{\kappa}^{(h)} \in L^1$ with

$$\hat{\kappa}^{(h)}(x - y) = \hat{k}^{(h)}(x, y) \tag{4.39}$$

for all $x, y \in \mathbb{R}^n$. $\qquad\square$

4.3.3   A Fredholm Criterion in $I + \mathcal{K}_f$

Having studied integral operators $K \in \mathcal{K}_f^\$$ a bit, we are now coming back to our original problem: a second kind integral equation

$$u + Ku = b$$

on BC. We write this equation as $Au = b$ with $A = I + K$. From Propositions 4.30 and 4.32 we know that $A \in \mathcal{A}_\$'$, and all its limit operators are of the form $A_h = I + K_h$; that is

$$(A_h u)(x) = u(x) + \int_{\mathbb{R}^n} \sum_{i=1}^{j} b_i^{(h)}(x)\, k_i\big(x - y, f^{(h)}(x), f^{(h)}(y)\big)\, \tilde{c}_i^{(h)}(y)\, u(y)\, dy$$

(4.40)

for all $u \in$ BC and $x \in \mathbb{R}^n$.

It is easily seen that $\mathcal{C}$ is contained in the ideal $\mathcal{J}$ (recall (4.13)) in $\mathcal{A}$ that is generated by convolution operators. Consequently, the equality $A = I + K$ with $K \in \mathcal{K}_f \subset \mathcal{C}$ is exactly the decomposition (4.14) of A into a multiplication operator and a locally compact part. Proposition 4.12 a) and our knowledge on the operator spectrum of A yield the following sufficient criterion for Fredholmness of A.

Proposition 4.34. *Let* $A = I + K$ *with* $K \in \mathcal{K}_f^\$$. *If all limit operators* (4.40) *of* A *are invertible on* L^∞, *then* A *is Fredholm on* L^∞ *and hence on* BC.

Proof. Since $\mathcal{K}_f^\$ \subset \mathcal{A}_\$'$, A is rich and band-dominated. Now Theorem 3.109 and Theorem 1 show that the invertibility of all limit operators of A on L^∞ implies that A is invertible at infinity. Finally, from $\mu(A) = I$ and Proposition 4.12 a) we get that A is Fredholm on L^∞ which clearly implies its Fredholmness on BC as well. $\square$

A much weaker result along these lines that, however, might be of interest for proving non-invertibility of A is the following: If $A = I + K$ with $K \in \mathcal{K}_f$ is invertible on L^∞, then all limit operators of A are invertible on L^∞.

4.3.4   The BC-FSM in $I + \mathcal{K}_f$

Since $K \in \mathcal{K}_f$ maps L^∞ into BC, we can, by Proposition 4.22, study the applicability of the BC-FSM (4.19) for $A = I + K$ by passing to its FSM (4.1) instead. In order to apply Theorem 4.2 on the finite section method, we restrict ourselves to operators on the axis, $n = 1$. By Theorem 4.2 we have to look at all operators of the form

$$QV_{-\tau} A_h V_\tau Q + P \qquad \text{with} \qquad A_h \in \sigma_+^{\mathrm{op}}(A) \qquad\qquad (4.41)$$

and

$$PV_{-\tau} A_h V_\tau P + Q \qquad \text{with} \qquad A_h \in \sigma_-^{\mathrm{op}}(A) \qquad\qquad (4.42)$$

with $\tau \in \mathbb{R}$. The operator (4.41) is invertible on L^∞ if and only if the operator that maps $u \in L^\infty$ to

$$u(x) + \int_{-\infty}^{0} \sum_{i=1}^{j} b_i^{(h)}(x-\tau) \, k_i\big(x-y, f^{(h)}(x-\tau), f^{(h)}(y-\tau)\big) \, \tilde{c}_i^{(h)}(y-\tau) \, u(y) \, dy \quad (4.43)$$

with $x < 0$ is invertible on the negative half axis, or, equivalently,

$$u(x) + \int_{-\infty}^{\tau} \sum_{i=1}^{j} b_i^{(h)}(x) \, k_i\big(x-y, f^{(h)}(x), f^{(h)}(y)\big) \, \tilde{c}_i^{(h)}(y) \, u(y) \, dy, \quad x < \tau$$

is invertible on the half axis $(-\infty, \tau)$ for the corresponding sequence $h = (h_1, h_2, \ldots)$ leading to a limit operator at plus infinity. And analogously, the operator (4.42) is invertible if and only if the operator that maps u to

$$u(x) + \int_{0}^{+\infty} \sum_{i=1}^{j} b_i^{(h)}(x-\tau) \, k_i\big(x-y, f^{(h)}(x-\tau), f^{(h)}(y-\tau)\big) \, \tilde{c}_i^{(h)}(y-\tau) \, u(y) \, dy \quad (4.44)$$

with $x > 0$ is invertible on the positive half axis, or, equivalently,

$$u(x) + \int_{\tau}^{+\infty} \sum_{i=1}^{j} b_i^{(h)}(x) \, k_i\big(x-y, f^{(h)}(x), f^{(h)}(y)\big) \, \tilde{c}_i^{(h)}(y) \, u(y) \, dy, \quad x > \tau$$

is invertible on the half axis $(\tau, +\infty)$ for the corresponding sequence $h = (h_1, h_2, \ldots)$ leading to a limit operator at minus infinity.

By Proposition 4.22, Theorem 4.2 and Remark 4.3 a), the modified finite section method is applicable to $A = I + K$ if and only if

- A is invertible on L^∞,

- for every sequence h leading to a limit operator at plus infinity, the set $\{(4.43)\}_{\tau \in \mathbb{R}}$ is essentially invertible on $L^\infty(-\infty, 0)$, and

- for every sequence h leading to a limit operator at minus infinity, the set $\{(4.44)\}_{\tau \in \mathbb{R}}$ is essentially invertible on $L^\infty(0, +\infty)$.

Remark 4.35. a) If $c_i \in BUC$ for all $i = 1, \ldots, j$, then both the operators (4.43) and (4.44) depend continuously on $\tau \in \mathbb{R}$ which is why 'essentially invertible' is the same as 'uniformly invertible' in the two statements above in this case.

b) If, as in Example 4.33, all of f, b_i and c_i are slowly oscillating, then we have $A_h = I + C_{F\hat\kappa^{(h)}}$ with $\hat\kappa^{(h)} \in L^1$ as introduced in (4.39) in Example 4.33. In this case, by Proposition 4.34, A is Fredholm if -1 is not in the spectrum of any $C_{F\hat\kappa^{(h)}}$; that is, all the (closed, connected) curves $F\hat\kappa^{(h)}(\dot{\mathbb{R}}) \subset \mathbb{C}$ stay away from the point -1. Moreover, the BC-FSM is applicable to A if and only if A is invertible and all curves $F\hat{\kappa}^{(h)}(\mathbb{R})$, on top of staying away from -1, have winding number zero with respect to this point.

c) In some cases (see Example 4.36 below) the functions $\hat{k}^{(h)}(x, y)$ from (4.38) in Example 4.33 even depend on $|x-y|$ only, which shows that the same is true for $\hat{\kappa}^{(h)}(x-y) := \hat{k}^{(h)}(x, y)$ then. If we then look at the applicability of the BC-FSM for $n = 1$, we get the following interesting result: The invertibility of A already implies the applicability of the BC-FSM. Indeed, if A is invertible, then, all limit operators A_h are invertible, which shows that all functions $F\hat{\kappa}_h$ stay away from the point -1. But from $F\hat{\kappa}^{(h)}(z) = F\hat{\kappa}^{(h)}(-z)$ for all $z \in \mathbb{R}$ we get that the point $F\hat{\kappa}^{(h)}(z)$ traces the same curve (just in opposite directions) for $z < 0$ and for $z > 0$. So the winding number of the curve $F\hat{\kappa}^{(h)}(\dot{\mathbb{R}})$ around -1 is automatically zero then.

Example 4.36. Let us come back to Example 4.25 where, as we found out earlier, $n = 1$, $j = 2$, $b_1 \equiv -1/\pi$, $c_1 = f'$, $b_2 \equiv 1/\pi$, $c_2 \equiv 1$,

$$k_1(r, s, t) = \frac{r}{r^2 + (s-t)^2} - \frac{r}{r^2 + (s+t)^2}$$

and

$$k_2(r, s, t) = \frac{s-t}{r^2 + (s-t)^2} + \frac{s+t}{r^2 + (s+t)^2}.$$

In addition, suppose that $f'(x) \to 0$ as $x \to \infty$. Then, by Lemma 3.45 b), all of b_1, b_2, c_1, c_2 and f are slowly oscillating, and, for every sequence h leading to infinity such that the strict limit $f^{(h)}$ exists, we have that $b_1^{(h)} \equiv -1/\pi$, $c_1^{(h)} \equiv 0$, $b_2^{(h)} \equiv 1/\pi$, $c_2^{(h)} \equiv 1$, and $f^{(h)} \geq f_- > 0$ is a constant function, whence

$$
\begin{aligned}
\hat{k}^{(h)}(x, y) &= \frac{1}{\pi}\left(\frac{f^{(h)} - f^{(h)}}{(x-y)^2 + (f^{(h)} - f^{(h)})^2} + \frac{f^{(h)} + f^{(h)}}{(x-y)^2 + (f^{(h)} + f^{(h)})^2}\right) \\
&= \frac{2f^{(h)}}{\pi}\frac{1}{(x-y)^2 + 4(f^{(h)})^2} =: \hat{\kappa}^{(h)}(x-y), \qquad x, y \in \mathbb{R}^n
\end{aligned}
$$

where $f^{(h)}$ is an accumulation value of f at infinity.

Now it remains to check the function values of the Fourier transform $F\hat{\kappa}^{(h)}$. A little exercise in contour integration shows that $F\hat{\kappa}^{(h)}(z) = \exp(-2f^{(h)}|z|)$ for $z \in \mathbb{R}$ (cf. Remark 4.35 c)). So $F\hat{\kappa}^{(h)}(\mathbb{R})$ stays away from -1 and has winding number zero.

Consequently, by our criteria derived earlier, we get that A is Fredholm and that the BC-FSM is applicable if and only if A is invertible. As discussed in [57], by other, somewhat related arguments, it can, in fact, be shown that A is invertible, even when f is not slowly oscillating. Precisely, injectivity of A can be established via applications of the maximum principle to the associated BVP, and then limit operator-type arguments can be used to establish surjectivity.

We note also that the modified version of the BC-FSM proposed in [19] could be applied in this case. (This method approximates the actual surface function f by an f for which f' is compactly supported before applying the finite section.) For this modified version the arguments of [19] and the invertibility of A establish applicability even when f is not slowly oscillating. □

4.4 Comments and References

There is a lot of ongoing research on the finite section method for band-dominated operators in general, as well as for special coefficients. STEFFEN ROCH has written a very nice paper [75] on this topic that reviews the state of the art for the case $p = 2$, $\mathbf{X} = \mathbb{C}$ including some new contributions to the theory. The finite section method for band-dominated operators with almost periodic coefficients has been studied by RABINOVICH, ROCH and SILBERMANN in [71]. The method that is called BC-FSM here is usually referred to as "the finite section method" in the literature dealing with operators $A = I + K$ on BC.

The first versions of Theorem 4.2 appeared in [67], [68] and [69]. The results for the continuous case and $\mathbb{D} = \mathbb{R}$ go back to [50] and [48]. Proposition 4.8 is the main result of [51] by RABINOVICH, ROCH and the author. Later, ROCH [74] generalized this result to band-dominated operators with slowly oscillating symbol. Most of Section 4.2 goes back to [50]. Parts of this, for example Subsection 4.2.3, were essentially already contained in [46].

Section 4.3 is recent work by CHANDLER-WILDE and the author [18]. For some particular problems of the discussed type, Fredholmness and even invertibility have been established before [2, 3, 4, 21, 22, 23]. Also note that a sufficient criterion for the applicability of a modified form of the finite section method for rough surface scattering was derived in [19]. We would, however, like to underline the generality of the approach presented here, as well as its very nice symbiosis with the limit operator concept.

Index

Bibliography

[1] G. R. ALLAN: Ideals of vector-valued functions, *Proc. London Math. Soc.*, **18** (1968), 193–216.

[2] T. ARENS: Uniqueness for Elastic Wave Scattering by Rough Surfaces, *SIAM J. Math. Anal.* **33** (2001), 461–476.

[3] T. ARENS: Existence of Solution in Elastic Wave Scattering by Unbounded Rough Surfaces, *Math. Meth. Appl. Sci.* **25** (2002), 507–528.

[4] T. ARENS, S. N. CHANDLER-WILDE and K. HASELOH: Solvability and spectral properties of integral equations on the real line. II. L^p-spaces and applications, *J. Integral Equations Appl.* **15** (2003), no. 1, 1–35.

[5] A. BÖTTCHER and S. M. GRUDSKY: *Spectral Properties of Banded Toeplitz Matrices*, SIAM, Philadelphia 2005.

[6] A. BÖTTCHER, YU. I. KARLOVICH and V. S. RABINOVICH: The method of limit operators for one-dimensional singular integrals with slowly oscillating data, *J. Operator Theory*, **43** (2000), 171–198.

[7] A. BÖTTCHER, YU. I. KARLOVICH and I. SPITKOVSKI: *Convolution Operators and Factorization of Almost Periodic Matrix Functions*, Birkhäuser Verlag, Basel 2002.

[8] A. BÖTTCHER and B. SILBERMANN: Infinite Toeplitz and Hankel matrices with operator-values entries., *SIAM J. Math. Anal.*, **27** (1996), 3, 805–822.

[9] A. BÖTTCHER and B. SILBERMANN: *Analysis of Toeplitz Operators*, Akademie-Verlag, Berlin, 1989 and Springer Verlag, Berlin, Heidelberg, New York 1990, 2nd edition: Springer Verlag, Berlin, Heidelberg, New York 2006.

[10] A. BÖTTCHER and B. SILBERMANN: *Introduction to large truncated Toeplitz Matrices*, Springer Verlag, New York 1999.

[11] P. C. BRESSLOFF and S. COOMBES: Mathematical reduction techniques for modelling biophysical neural networks, *In: Biophysical Neural Networks*, Ed. Roman R Poznanski, Gordon and Breach Scientific Publishers, 2001.

[12] R. C. BUCK: Bounded Continuous Functions on a Locally Compact Space, *Michigan Math. J.* **5** (1958), 95–104.

[13] C. CORDUNEANU: *Almost periodic functions*, Chelsea Publishing Company, New York, 1989.

[14] C. H. CHAN, K. PAK and H. SANGANI: Monte-Carlo simulations of large-scale problems of random rough surface scattering and applications to grazing incidence, *IEEE Trans. Anten. Prop.*, **43**, 851–859.

[15] S. N. CHANDLER-WILDE: Some uniform stability and convergence results for integral equations on the real line and projection methods for their solution, *IMA J. Numer. Anal.* **13** (1993), no. 4, 509–535.

[16] S. N. CHANDLER-WILDE, S. LANGDON and L. RITTER: A high-wavenumber boundary-element method for an acoustic scattering problem, *Phil. Trans. R. Soc. Lond. A.*, **362** (2004), 647–671.

[17] S. N. CHANDLER-WILDE and M. LINDNER: Generalized Collective Compactness and Limit Operators, submitted to *Integral Equations Operator Theory* (2006).

[18] S. N. CHANDLER-WILDE and M. LINDNER: Boundary integral equations on unbounded rough surfaces: Fredholmness and the Finite Section Method, submitted for publication (2006).

[19] S. N. CHANDLER-WILDE and A. MEIER: On the stability and convergence of the finite section method for integral equation formulations of rough surface scattering, *Math. Methods Appl. Sci.*, **24** (2001), no. 4, 209–232.

[20] S. N. CHANDLER-WILDE, M. RAHMAN and C. R. ROSS: A fast two-grid and finite section method for a class of integral equations on the real line with application to an acoustic scattering problem in the half-plane, *Numerische Mathematik*, **93** (2002), no. 1, 1–51.

[21] S. N. CHANDLER-WILDE, C. R. ROSS and B. ZHANG: Scattering by infinite one-dimensional rough surfaces, *Proc. R. Soc. Lond. A.*, **455** (1999), 3767–3787.

[22] S. N. CHANDLER-WILDE and C. R. ROSS: Scattering by rough surfaces: The Dirichlet problem for the Helmholtz equation in a non-locally perturbed half-plane, *Math. Meth. Appl. Sci.*, **19** (1996), 959–976.

[23] S. N. CHANDLER-WILDE and B. ZHANG: On the solvability of a class of second kind integral equations on unbounded domains, *J. Math. Anal. Appl.* **214** (1997), no. 2, 482–502.

[24] S. N. CHANDLER-WILDE and B. ZHANG: A generalised collectively compact operator theory with an application to second kind integral equations on unbounded domains, *J. Integral Equations Appl.* **14** (2002), 11–52.

[25] D. COLTON and R. KRESS: *Integral Equation Methods in Scattering Theory*, Wiley, New York 1983.

[26] E. B. DAVIES: Spectral Theory of Pseudo-Ergodic Operators, *Commun. Math. Phys.* **216** (2001), 687–704.

[27] R. G. DOUGLAS: *Banach Algebra Techniques in Operator Theory*, Academic Press, New York, London 1972.

[28] J. A. FAVARD: Sur les equations differentiales a coefficients presque periodiques, *Acta Math.* **51** (1927), 31–81.

[29] I. GOHBERG and I. A. FELDMAN: *Convolution equations and projection methods for their solutions*, Nauka, Moskva 1971 (Russian; Engl. translation: Amer. Math. Soc. Transl. of Math. Monographs 41, Providence, R.I. 1974).

[30] I. GOHBERG and M. G. KREIN: Systems of integral equations on the semi-axis with kernels depending on the difference of arguments, *Usp. Mat. Nauk* **13** (1958), no. 5, 3–72 (Russian).

[31] I. GOHBERG and N. KRUPNIK: *One-dimensional linear singular integral operators*, Vol. I., Birkhäuser Verlag, Basel, Boston, Stuttgart 1992.

[32] M. B. GORODETSKI: On the Fredholm theory and the finite section method for multidimensional discrete convolutions, *Sov. Math.* **25** (1981), no. 4, 9–12.

[33] A. GROTHENDIECK: Une Caracterisation Vectorielle-Metrique de Espaces L^1, *Canad. Math. J.* **7** (1955), 552–561.

[34] W. HACKBUSCH: *Integral Equations*, Birkhäuser Verlag, 1995.

[35] R. HAGEN, S. ROCH and B. SILBERMANN: *Spectral Theory of Approximation Methods for Convolution Equations*, Birkhäuser Verlag, 1995.

[36] R. HAGEN, S. ROCH and B. SILBERMANN: $C^*-Algebras$ *and Numerical Analysis*, Marcel Dekker, Inc., New York, Basel, 2001.

[37] P. R. HALMOS: Ten problems in Hilbert space, *Bull. Amer. Math. Soc.* **76** (1970), 887–933.

[38] N. HATANO and D. R. NELSON: Vortex pinning and non-Hermitian quantum mechanics. *Phys. Rev. B* **77** (1996), 8651–8673.

[39] Y. KATZNELSON: *An Introduction to Harmonic Analysis*, John Wiley 1968, Dover Publications 1976 and Cambridge University Press 2004.

[40] M. G. KREIN: Integral equations on the semi-axis with kernels depending on the difference of arguments, *Usp. Mat. Nauk* **13** (1958), no. 2, 3–120 (Russian).

[41] V. G. KURBATOV: *Functional Differential Operators and Equations*, Kluwer Academic Publishers, Dordrecht, Boston, London 1999.

[42] S. LANGDON and S. N. CHANDLER-WILDE: A wavenumber independent boundary element method for an acoustic scattering problem, *Isaac Newton Institute for Mathematical Sciences*, Preprint NI03049-CPD, to appear in *SIAM J. Numer. Anal.*

[43] B. V. LANGE and V. S. RABINOVICH: On the Noether property of multidimensional discrete convolutions, *Mat. Zam.* **37** (1985), no. 3, 407–421.

[44] B. V. LANGE and V. S. RABINOVICH: On the Noether property of multidimensional operators of convolution type with measurable bounded coefficients, *Izv. Vyssh. Uchebn. Zaved., Mat.* **6** (1985), 22–30 (Russian).

[45] B. V. LANGE and V. S. RABINOVICH: Pseudo-differential operators on $\mathbb{R}^n$ and limit operators, *Mat. Sbornik* **129** (1986), no. 2, 175–185 (Russian, English transl. *Math. USSR Sbornik* **577** (1987), no. 1, 183–194).

[46] M. LINDNER and B. SILBERMANN: Finite Sections in an Algebra of Convolution and Multiplication Operators on $L^\infty(\mathbb{R})$, *TU Chemnitz, Preprint 6* (2000).

[47] M. LINDNER and B. SILBERMANN: Invertibility at infinity of band-dominated operators in the space of essentially bounded functions, *In: Operator Theory – Advances and Applications* **147**, *The Erhard Meister Memorial Volume*, Birkhäuser 2003, 295–323.

[48] M. LINDNER: The finite section method in the space of essentially bounded functions: An approach using limit operators, *Numer. Func. Anal. & Optim.* **24** (2003) no. 7&8, 863–893.

[49] M. LINDNER: Classes of Multiplication Operators and Their Limit Operators, *Journal for Analysis and its Applications* **23** (2004), no. 1, 187–204.

[50] M. LINDNER: Limit Operators and Applications on the Space of essentially bounded Functions, *Dissertation, TU Chemnitz* 2003.

[51] M. LINDNER, V. S. RABINOVICH and S. ROCH: Finite sections of band operators with slowly oscilating coefficients, *Linear Algebra and Applications* **390** (2004), 19–26.

[52] X. LIU, G. STRANG and S. OTT: Localized eigenvectors from widely spaced matrix modifications. *SIAM J. Discr. Math.* **16** (2003), no. 3, 479–498.

[53] W. C. H. MCLEAN: *Strongly Elliptic Systems and Boundary Integral Equations*, Cambridge University Press 2000.

[54] E. M. MUHAMADIEV: On normal solvability and Noether property of elliptic operators in spacesof functions on $\mathbb{R}^n$, Part I: *Zapiski nauchnih sem. LOMI* **110** (1981), 120–140 (Russian).

[55] E. M. MUHAMADIEV: On normal solvability and Noether property of elliptic operators in spacesof functions on $\mathbb{R}^n$, Part II: *Zapiski nauchnih sem. LOMI* **138** (1985), 108–126 (Russian).

[56] F. NOETHER: Über eine Klasse singulärer Integralgleichungen, *Math. Ann.* **82** (1921), 42–63.

[57] M. D. PRESTON, P. G. CHAMBERLAIN and S. N. CHANDLER-WILDE: An integral equation method for a boundary value problem arising in unsteady water wave problems, *Proceedings of the UK BIM5, Liverpool*, September 2005.

[58] S. PRÖSSDORF and B. SILBERMANN: *Numerical Analysis for Integral and Related Operator Equations*, Akademie-Verlag, Berlin, 1991 and Birkhäuser Verlag, Basel, Boston, Berlin 1991.

[59] V. S. RABINOVICH: Fredholmness of pseudo-differential operators on $\mathbb{R}^n$ in the scale of $L_{p,q}$-spaces, *Sib. Mat. Zh.* **29** (1988), no. 4, 635–646 (Russian, English transl. *Sib. Math. J.* **29** (1998), no. 4, 635–646).

[60] V. S. RABINOVICH: Criterion for local invertibility of pseudo-differential operators with operator symbols and some applications, *In:* Proceedings of the St. Petersburg Mathematical Society, **V**, *AMS Transl. Series 2*, **193** (1998), 239–260.

[61] V. S. RABINOVICH and S. ROCH: Local theory of the Fredholmness of band-dominated operators with slowly oscillating coefficients, *Toeplitz matrices and singular integral equations (Pobershau, 2001)*, 267–291, Oper. Theory Adv. Appl., 135, Birkhäuser, Basel, 2002.

[62] V. S. RABINOVICH and S. ROCH: An axiomatic approach to the limit operators method, *In:* Singular integral operators, factorization and applications, Operator Theory: Advances and Applications, **142**, 263–285, Birkhäuser, Basel, 2003.

[63] V. S. RABINOVICH and S. ROCH: Algebras of approximation sequences: Spectral and pseudo-spectral approximation of band-dominated operators, *Linear algebra, numerical functional analysis and wavelet analysis*, 167–188, Allied Publ., New Delhi, 2003.

[64] V. S. RABINOVICH and S. ROCH: The essential spectrum of Schrödinger operators on lattices, *Preprint Nr. 2443, Technical University Darmstadt*, 2006 (to appear in *Journal of Physics, Ser. A*, 2006).

[65] V. S. RABINOVICH and S. ROCH: Reconstruction of input signals in time-varying filters, *Preprint Nr. 2418, Technical University Darmstadt*, 2006 (to appear in *Numer. Func. Anal. & Optim.*, 2006).

[66] V. S. RABINOVICH, S. ROCH and J. ROE: Fredholm indices of band-dominated operators, *Integral Equations Operator Theory* **49** (2004), no. 2, 221–238.

[67] V. S. RABINOVICH, S. ROCH and B. SILBERMANN: Fredholm Theory and Finite Section Method for Band-dominated operators, *Integral Equations Operator Theory* **30** (1998), no. 4, 452–495.

[68] V. S. RABINOVICH, S. ROCH and B. SILBERMANN: Band-dominated operators with operator-valued coefficients, their Fredholm properties and finite sections, *Integral Equations Operator Theory* **40** (2001), no. 3, 342–381.

[69] V. S. RABINOVICH, S. ROCH and B. SILBERMANN: Algebras of approximation sequences: Finite sections of band-dominated operators, *Acta Appl. Math.* **65** (2001), 315–332.

[70] V. S. RABINOVICH, S. ROCH and B. SILBERMANN: *Limit Operators and Their Applications in Operator Theory*, Operator Theory: Advances and Applications, **150**, Birkhäuser Basel 2004.

[71] V. S. RABINOVICH, S. ROCH and B. SILBERMANN: Finite Sections of band-dominated operators with almost periodic coefficients, *Preprint Nr. 2398, Technical University Darmstadt*, 2005 (to appear in *Modern Operator Theory and Applications*, **170**, *The Simonenko Anniversary Volume*, Birkhäuser 2006).

[72] M. REED and B. SIMON: *Methods of Modern Mathematical Physics, II. Fourier Analysis, Self-Adjointness*, Academic Press, New York, San Francisco, London 1975.

[73] J. ROE: Band-dominated Fredholm operators on discrete groups, *Integral Equations Operator Theory* **51** (2005), no. 3, 411–416.

[74] S. ROCH: Band-dominated operators on ℓ^p−spaces: Fredholm indices and finite sections, *Acta Sci. Math.* **70** (2004), no. 3–4, 783–797.

[75] S. ROCH: Finite sections of band-dominated operators, *Preprint Nr. 2355, Technical University Darmstadt*, 2004 (to appear in *Memoirs AMS*, 2006).

[76] S. ROCH and B. SILBERMANN: Non-strongly converging approximation methods, *Demonstratio Math.* **22** (1989), no. 3, 651–676.

[77] S. SAKAI: C^*-*algebras and* W^*-*algebras*, Springer Berlin, Heidelberg, New York 1971.

[78] D. SARASON: Toeplitz Operators with semi-almost periodic symbols, *Duke Math. J.* **44**, no. 2 (1977), 357-364.

[79] S. Sauter and C. SCHWAB: *Randelementmethoden*, Teubner 2004.

[80] H. H. SCHAEFER: *Topological vector spaces*, MacMillan Company New York, Collier-MacMillan Ltd., London 1966.

[81] B. YA. SHTEINBERG: Operators of convolution type on locally compact groups, *Rostov-na-Donu, Dep. at VINITI* (1979), 715–780 (Russian).

[82] I. B. SIMONENKO: Operators of convolution type in cones, *Mat. Sb.* **74** (116),(1967), 298–313 (Russian).

[83] I. B. SIMONENKO: On multidimensional discrete convolutions, *Mat. Issled.* **3** (1968), no. 1, 108–127 (Russian).

[84] E. M. STEIN and G. WEISS: *Introduction to Fourier Analysis on Euclidean Spaces*, Princeton University Press, 1971.

[85] L. N. TREFETHEN and M. EMBREE: *Spectra and Pseudospectra*, Princeton University Press, 2005.

[86] L. TSANG, C. H. CHAN, H. SANGANI, A. ISHIMARU and P. PHU: A banded matrix iterative approach to Monte-Carlo simulations of large-scale random rough surface scattering, *TE case, J. Electromagnetic Waves Appl.*, **7** (1993), 1185–1200.

[87] N. WIENER and E. HOPF: Über eine Klasse singulärer Integralgleichungen, *Sitzungsberichte Akad. Wiss. Berlin* (1931), 696–706.

[88] K. YOSIDA: *Functional Analysis*, Springer 1995.

Frontiers in Mathematics

This new series is designed to be a repository for up-to-date research results which have been prepared for a wider audience. Graduates and postgraduates as well as scientists will benefit from the latest developments at the research frontiers in mathematics and at the "frontiers" between mathematics and other fields like computer science, physics, biology, economics, finance, etc. All volumes will be online available at SpringerLink.

■ **Bouchut, F.**, CNRS & Ecole Normale Sup., Paris, France

Nonlinear Stability of Finite Volume Methods for Hyperbolic Conservation Laws and Well-Balanced Schemes for Sources

2004. 142 pages. Softcover. ISBN 3-7643-6665-6

■ **Clark, J.**, Otago Univ., New Zealand / **Lomp, C.**, Univ. di Porto, Portugal / **Vanaja, N.**, Mumbai Univ., India / **Wisbauer, R.**, Univ. Düsseldorf, Germany

Lifting Modules

2006. 408 pages. Softcover. ISBN 3-7643-7572-8

■ **De Bruyn, B.**, Ghent University, Ghent, Belgium

Near Polygons

2006. 276 pages. Softcover. ISBN 3-7643-7552-3

■ **Henrot, A.**, Université Henri Poincaré, Vandoeuvre-les-Nancy, France

Extremum Problems for Eigenvalues of Elliptic Operators

2006. 216 pages. Softcover. ISBN 3-7643-7705-4

■ **Kasch, F.**, Universität München, Germany / **Mader, A.**, Hawaii University

Rings, Modules, and the Total

2004. 148 pages. Softcover. ISBN 3-7643-7125-0

■ **Krausshar, R.S.**, Ghent University, Ghent, Belgium

Generalized Analytic Automorphic Forms in Hypercomplex Spaces

2004. 182 pages. Softcover. ISBN 3-7643-7059-9

■ **Thas, K.**, Ghent University, Ghent, Belgium

Symmetry in Finite Generalized Quadrangles

2004. 240 pages. Softcover. ISBN 3-7643-6158-1

■ **Zaharopol, R.**, Math. Reviews, Ann Arbor, USA

Invariant Probabilities of Markov-Feller Operators and Their Supports

2005. 120 pages. Softcover. ISBN 3-7643-7134-X